Management Division

Presentations at the 2010 AIChE Annual Meeting

Salt Lake City, Utah, USA
7-12 November 2010

ISBN: 978-1-61782-151-6

Printed from e-media with permission by:

Curran Associates, Inc.
57 Morehouse Lane
Red Hook, NY 12571

proceedings.com

Some format issues inherent in the e-media version may also appear in this print version.

Printed by Curran Associates, Inc. (2011)

AIChE
3 Park Avenue
New York, NY 10016-5991

Phone: (203) 702-7660
Fax: (203) 775-5177

www.aiche.org

Additional copies of this publication are available from:

Curran Associates, Inc.
57 Morehouse Lane
Red Hook, NY 12571 USA
Phone: 845-758-0400
Fax: 845-758-2634
Email: curran@proceedings.com
Web: www.proceedings.com

TABLE OF CONTENTS

(118a) Project Management in a Matrix Organization

Frank van Lier, The Lubrizol Corporation, Wickliffe, OH

Successful project management is a critical aspect for organizational success. It becomes even more challenging when the project manager and team members are part of a matrix organization. The ability to gain buy-in on goals and objectives, access the needed people and resources, and carry out projects requires deft people skills. Benefits can include higher quality outcomes and increased implementation speeds. This presentation will cover one industry veteran's perspective on how to accomplish project objectives in a highly matrixed organization.

See more of this Session: Project Management: What's New? What Works?
See more of this Group/Topical: Management Division

(118b) Environmental Law 101 Module

Chris A. Paul, Mcafee & Taft, Tulsa, OK

This tutorial, Environmental Law 101 Module, provides a summary of legal issues for engineering professionals and organizations most relevant for new project managers. An extended abstract will be submitted prior to the conference.

See more of this Session: Project Management: What's New? What Works?
See more of this Group/Topical: Management Division

Planning, Managing, and Best Practices for Project Procurement

Dr. Eldon R. Larsen, Marshall University Graduate College, South Charleston, WV

Paul Jason Dean, Kentucky Transportation, Ashland, KY

Abstract

Project procurement is all about acquiring materials, services, equipment, supplies, and anything else outside of the project team that is needed to complete a project satisfactorily. Procurement of goods and services can be one of the costliest portions of a project's budget. Procurement can also be a significant source of project risk and schedule delay. Yet, it is often underestimated in importance to the project. In this paper we discuss various aspects of project procurement, including contract management, and present some best practices.

Procurement Planning

Before a project team starts into actual project procurement, it is important to plan how procurement will take place, what procedures will be followed, what documentation will be needed, what levels of approval and signatures will be required, and how all of this affects the rest of the project planning, risks, resources, and so on. Additionally, the project team needs to decide what will be purchased, versus what the team will provide. Procurement involves both buyers and sellers. In fact, the project team could be a buyer, a seller, or both.

Often, procurement planning will include a determination of potential sellers and requirements for those sellers. Some organizations have lists of pre-qualified and/or preferred sellers to whom they specifically go for goods and services. Other organizations are required to openly solicit bids from anyone interested in providing the needed good or service. Thus, it is important to understand what policies will govern any given project in your organization.

When planning project procurements, there are many factors to consider:

- Project or product scope constraints related to performance, timing, and cost
- Health, safety, and environmental requirements
- Security concerns
- Intellectual property concerns
- Legal issues surrounding licenses, permits, regulations, safety, and employment
- Partnership arrangements or teaming agreements
- Organizational procurement policies and procedures
- Previous experience with sellers
- Lessons learned regarding procurements for previous projects
- Insurance needs
- Roles and responsibilities
- Past and present information about potential sellers

- Unique requirements for the project
- Previously established agreements and contracts
- Make-or-buy decisions
- How to manage multiple sellers, where applicable
- Important procurement lead times needed in the project plan
- Needed procurement completion dates
- The potential for needed insurance contracts and performance bonds
- Documentation that will be required of contractors by the buyer before, during, and at completion of contract work
- Seller selection criteria, including making these public when required
- Qualified sellers lists giving the names of potential sellers who have already been identified as qualified to provide a specific good or service

It is also important to look at whether it is better to make or to buy some goods needed for the project, and whether to supply internally, or to buy from outside, some needed services. Sometimes, rather than a make-or-buy decision, a project manager could also be faced with a *rent-or-buy* decision, similar to a make-or-buy decision, but where something is potentially rented rather than purchased.

Sometimes procurement planning will involve determining whether the buyer wants to use a *sole source* or a *single source.* Though these terms seem very similar, there is an important technical difference. A *sole source* is used when a specific seller is the only one who can provide exactly what the buyer needs, perhaps due to intellectual property protection like patents. In contrast, a *single source* might be used because of preferences of the buyer due to factors like negotiated long-term special pricing or special qualifications or performance history and a desire to maintain long-term relationships between the buyer and seller.

The result of procurement planning is a plan that guides how the procurement process will be managed. This plan can range from somewhat informal to very detailed and very formalized, depending upon the size and the needs of the project and according to organizational policies. Once procurement planning has occurred, it is important to go back an integrate this planning into the main plan for the project and into the work breakdown structure, network diagram, and Gantt chart as needed.

It is important in the procurement planning to clearly define roles and responsibilities for people involved in the procurement process. In larger projects it is highly recommended that an owner's technical representative be appointed with clearly defined responsibilities in monitoring project progress, quality, billing, and so forth as deemed appropriate for the project—many inexperienced project managers regret later on that they did not pay enough attention to this role.

Also, an important product of procurement planning should be a decision as to what kind of contracts, if any, will be used and how bidding on those contracts is to take place. Also, if independent estimates are to be obtained for the contract work, the procurement planning process should determine how this is to be obtained.

Because procurement often involves obtaining or providing project-needed goods and services through contracts and purchase orders, there is a need for project managers to understand contract administration and various purchasing terms and procedures.

Contracts

Contracts are legal, mutually binding agreements between buyers and sellers. When a buyer enters into a contract, he is legally bound to provide the monies agreed-upon for the sellers' products. Because contracts are legally binding, disagreements can be subject to resolution in the courts, if needed. Sometimes contracts can also take the form of *agreements, understandings,* or *purchase orders*.

Sometimes project managers decide to enter into contracts with sellers in order to share project risks with a contractor, when the project manager does not want to absorb the full risk, or when the project team is not adequately qualified or able to provide a needed good or service.

It is important to write project contracts so that they meet the specific needs of the project. Even if projects seem to be very similar, each project is still unique in some way. Thus, each contract can vary even for similar projects.

Who writes the contract—the buyer or the seller? In practice it could be either, depending upon the situation. However, if someone outside of the project team is going to write the contract, it is strongly advisable for the project manager to draft what he or she thinks should be in the contract, before seeing what the other party writes. This will help the project manager (1) to think about what is really needed and (2) not to become so focused only on what the other party writes that he forgets to include other important aspects that should be included in the contract. The party who writes the contract often will write it to be in his favor, so caution is important. Take time to get the contract right.

Contracts need to be approved and signed within an organization's policies, so the project manager needs to become familiar with those policies. Additionally, it is good practice to become very familiar with one's purchasing department and contract administrators in organizations where those will be called upon for help.

There are many types of contracts, some beyond the scope of this book. The more common, standard contract types are discussed below. Keep in mind that not all contract types are appropriate for all organizations. Organizational policies could greatly restrict what contract options are allowable. For example, in many governmental organizations there are laws that specify what types of contracts may be used.

Fixed-Price or Lump-Sum Contracts

Fixed-price or lump-sum contracts are those in which the total set price is agreed upon for a clearly defined good or service. The buyer must clearly define what good or service is desired. The seller is legally bound to provide the good or service for the fixed price. If there are mutually agreed-upon changes to the project scope, then the price can be adjusted accordingly. The simplest form of a fixed-price contract is a purchase order for a product to be provided by a certain date for a certain price. However, there are several variations of fixed-price contracts, and some of them can include incentives to meet specific project objectives. A few of these variations follow.

Firm Fixed Price Contracts (FFP). FFP contracts are one of the most commonly used project contracts. People like these contracts because the price is clearly agreed upon and understood. Any increased costs will be completely born by the seller, unless there is a mutually agreed-upon scope change, in which case the resulting price changes are born by the buyer.

Fixed Price Incentive Fee Contracts (FPIF). FPIF contracts are similar to FFP contracts, *except* that FPIF contracts allow for performance variations with incentives to meet pre-defined performance measures of success. These measures of success can and should be related to cost, schedule, and project and product performance. The final price is not known until the project is completed. There can be pre-defined price ceilings, above which the seller bears any additional cost and must still complete the contract obligations.

Fixed Price with Economic Price Adjustment (FP-EPA). FP-EPA contracts are often used when the contract spans a long period of time, like several years, which can affect the base cost of raw materials and labor costs. For example, over several years the prices for steel, concrete, and lumber can vary greatly, making it extremely difficult for a contractor to accurately estimate long-term costs of building materials. Therefore, FP-EPA contracts provide for pre-defined final price adjustments due to changes in factors like material costs, inflation, or regulatory changes. These contracts should use reliable financial indices and should be used to protect both the buyer and the seller.

Cost Reimbursable Contracts

Cost reimbursable contracts reimburse the seller for his costs, plus a fee to allow the seller to make a profit. The sellers costs often include both direct and indirect or overhead costs, and the indirect costs are usually but not always some percentage of the direct costs. Additionally, there can be incentives to meet specific project objectives. There are several common variations in cost reimbursable contracts, some of which follow.

Cost Plus Fixed Fee Contracts (CPFF), also called Cost Plus Percentage of Cost Contracts (CPPC). CPFF and CPPC contracts are used when the buyer and seller agree that the seller is to be reimbursed for allowable costs, which are not fixed in advance, plus the seller is to receive a fee based upon a percentage of the estimated cost when the contract is initially signed, which might not be the final actual cost. The fee is paid at the end of the job and does not normally depend on other factors like contractor performance. If the project scope changes through an approved change control process, then the fee could change.

Cost Plus Incentive Fee Contracts (CPIF). In some ways the CPIF contract is similar to the CPFF contract, except that for the CPIF contract the contractor or seller receives a fee called an incentive bonus, which bonus is based upon meeting certain specified performance criteria in the contract. The performance criteria could involve cost metrics, schedule metrics, and product and project scope quality and performance metrics. For these contracts, if the seller is able to provide the required product or service for a final actual cost that is less than originally estimated when the contract was signed, then there is often some provision for both the seller and the buy to share in the savings, and thus there is an incentive for the seller to pay careful attention to costs. There could also be monetary incentives for the seller if the contract is completed early, but not always. Every project's contract will be different in what incentives are specified.

Cost Plus Award Fee Contracts (CPAF). CPAF contracts are very similar to CPIF contracts. The buyer reimburses the seller for allowable costs. The seller's fee is based upon meeting agreed-upon subjective measures of performance specified in the contract, and is usually not subject to appeal.

Time and Materials (T & M) Contracts

Time and materials contracts tend to be a kind of hybrid between fixed priced contracts and cost reimbursable contracts. Similar to cost reimbursable contracts, time and materials contracts are open-ended in that the actual allowable costs to be reimbursed could be more or less than the original estimated cost. Sometimes these contracts specify fixed-prices per unit of a good or service that will be reimbursed. Often there are *not-to-exceed* limits on how much will be reimbursed.

Common Procurement Documents

There are a number of important documents frequently involved in procurement activities. Some of these documents have several different names, depending in part on the industry or organization using them. Depending on the needs of the project, the procurement documents should provide information in a manner that provides pertinent information and helps achieve the procurement objectives as easily and effectively as possible. Standardized formats, checklists, clear statements of work, clear expectations, complete legal provisions and constraints are all useful and important. Sometimes public disclosure and advertising are either required or recommended, in order to be fair to all participants. Generally sellers should follow buyers' document guidelines. Below, some of the more common documents are discussed.

Request for Quotation (RFQ). Requests for quotations often are used to ask for small-dollar amounts or off-the-shelf commodities not needing much or any customization. Usually RFQs do not require negotiation, unless it is minimal.

Request for Information (RFI). Requests for information ask for information about a potential seller that would help the buyer to understand that seller's qualifications to do the work needed in a procurement.

Request for Proposal (RFP). Requests for proposals are sometimes called *solicitations* or a *request for tender* (RFT). The seller's response would be a *proposal.* RFPs generally ask for proposals for customized goods or services which are sometimes high-priced or extensive in nature. The seller's proposal should include relatively detailed information about the approach the seller would use to provide the needed good or service. RFPs can result in negotiations with respect to schedule, costs, and performance or scope or the work to be accomplished.

Invitation for Bid (IFB). Invitations for bids or IFB are sometimes called *requests for bids* or RFB. These documents are similar to RFPs, but the IFB terminology is more common for government projects' sealed bidding.

Letter of Intent. A letter of intent is simply a document prepared by the buyer stating the intent of the buyer to purchase a good or service from a specific seller.

Procurement can involve many other documents. The project manager, buyers, and potential sellers should all become familiar with the specific documents and procedures required by the particular buyer's organization.

Bidder Conferences

For large project needs, where multiple potential sellers will be bidding on work to be performed for a buyer, the buy might hold a bidder conference. Bidder conferences might also be called contractor conferences, vendor conferences, or pre-bid conferences. These meetings occur will all the bidders interested in bidding, generally at the same time and location, *prior to* any actual bidding.

These conferences are used for the buyer to present pertinent information to all the interested potential bidders. The information would include clarification on the specific requirements, expectations, and constraints of the work to be provided to the buyer. These meetings provide an opportunity for bidders to ask clarifying questions and to better understand the buyer's needs. *All* interested bidders should be given exactly the same information and opportunities to obtain specifications and clarifying information.

It is highly recommended that the buyer provide each interested bidder with a complete and detailed record of the bidder conference, including answers to questions, specifications and drawings, handouts, presentation slides, and so forth. The buyer might even consider providing each interested bidder with a videotape of the meeting. Most important is that each bidder is treated equally.

It is also recommended that the buyer provide the bidders with a clear understanding of the evaluation criteria and procedures that will be used to select the final bidder which will provide the needed work to the seller.

Independent Estimates

For potentially large, expensive contracts, buyers often asked a qualified independent agent or organization to make an independent estimate of what the contract work *should cost*. Sometimes this independent estimate is done internally by the buyer. The independent estimate is used as a kind of check on the bidders' proposals. Significant differences between the independent estimate and bidders' proposals might indicate that there is a misunderstanding of what the buyer wants, and there was not sufficient information given to the bidders. If a long period of time has elapsed between the time of the independent estimate and the bidders' proposals, inflation or other economic factors might be affecting the discrepancies.

Contract Negotiations

For many contracts, especially those where the work is extensive and expensive, once a specific seller has been selected by the buyer, they will enter into contract negotiations. The objective is to reach mutual agreement on all the details of the contract between them: what will be provided, when, the price, payment arrangements, how conflicts and disputes will be resolved, roles and responsibilities, authorities, terms and laws, rights to intellectual property, and criteria and procedures for contract closure. Additionally, and very importantly, the contract needs to make clear the procedures to follow for any changes to contract scope, schedule, or costs.

For small contracts, there might not be any contract negation at all. However, it is advisable for the project manager to check with legal counsel or procurement experts to help determine when negotiations are needed and how they are to take place.

Contract Closeout

Contract closeout is done to formally signify that the work of the contract has been accepted by the buyer and that the terms of the contract have been achieved. This procedure verifies that all contract deliverables have been delivered in an acceptable manner. Additionally, all contract documents are finalized, recorded, and archived for future reference. Contracts can also be closed out early if for some reason it is necessary to end to contract before the contract work has been completed. This obviously might result in a contract dispute and a negotiated settlement of the contract, the procedure for which should have been detailed in the contract itself.

Green Project Procurement

Green Procurement is a new and emerging trend that focuses on the environmental sustainability of a product. The focus on green products is to reduce carbon dioxide emissions, increase water efficiency, increase energy savings, improve indoor air quality and achieve an overall smaller environmental footprint.

Many government entities and companies are enacting green procurement activities with a view of minimizing their overall environmental impact, and with the ultimate goal of specifying environmental preference for a product that truly has credibility as being green. A claim that an environmental product is green is simply not justification to mandate the product into a project. A green product has to be as good as or better than the traditional alternative. When specifying products many organizations and developers initially analyze the upfront or initial cost of acquiring a product. A new way of thinking must be utilized where the overall life cycle cost and the environmental footprint of the product is considered. The planning phase of project development offers the best opportunity to investigate and specify green products. This is where alternatives can be identified and where the environmental technical standards are outlined, along with performance and functional requirements.

The U.S. Green Building Council has a program called LEED (Leadership in Energy and Environmental Design) Green Building Rating System and accelerates the global adoption of sustainable green building and development practices through the creation and implementation of universally understood and accepted tools and performance criteria.

E-procurement

E-procurement is the utilization of the internet to facilitate purchase and sale of supplies, work and services through the Internet. E-procurement is a tool that allows a user to reach many more suppliers/contractors that a project sponsor would not be able to reach in traditional publications. Government, private and trade association web sites allow the buyers and sellers of goods and services to view all available opportunities for bid invitations. E-procurement is heavily utilized in large supply chain industries or just in time manufacturing facilities.

Many agencies, suppliers and contractors utilize their own e-procurement web pages or that of an industry trade group to promote services, proposal, and opportunities in a manner that allows them to reach more users. The federal government, states, and many large companies have web pages that outline procurement opportunities, policies, procedures, rules, regulations, and frequently asked questions tabs which outline material relevant to procurement in their organizations. This is all in an effort to disseminate information to their respective audiences.

Procurement Best Practices

Before completing this chapter on procurement, let's look at some of the *best practices* in the procurement field. The following best-practices summary is based mainly on work by William Gindlesperger (2009) and by Mark Smith (2003).

- Establish common procurement definitions, and the organization's standardized contract clauses that can be adapted as needed by specific project needs
- Create and deploy an integrated contract management system with a complete record and audit trail; where possible make it web-based
- Identify needs for new contracts and sources of alternative supply quickly
- Identify risks and potential exposure, due to contracts
- Create a pre-qualified supplier pool in which all suppliers are uniformly and objectively evaluated—know the suppliers in the pool and maintain relationships with fewer, yet higher performing suppliers
- Create purchase order templates pre-populated with approved suppliers and supplier profile data
- Maintain, update and utilize software packages (internal and external) that outline average unit bid prices broken down by quantity, county, state, region and the United States, to allow for more accurate pricing and competitive bidding
- Create thorough bid specifications with full details before an invitation for bid
- Create a fair bidding environment and match specifications with capable suppliers
- Streamline procurement procedures and create efficiencies to save time and money; facilitate access to approved suppliers
- Create mutual acceptance of the process that every supplier understands
- Call for, and award bids in a timely manner
- Teach suppliers that past *performance* is important to receive contract opportunities, and then *price* is the essential to win those contracts
- Maintain and nurture positive relationships, with contractors, trade organizations, government entities, utility companies, suppliers, etc, as a means to reach out and ask questions and utilize their expertise when a problem or question arises
- Be active and participate in partnering – make it a plan for success, as this allows for solutions and answers to problems in a timely manner and may resolve many issues where future litigation may arise
- Adhere to and practice a code of ethics
- Establish transparency by sending bid results to all suppliers that submitted a bid
- Monitor every step of the workflow process—ensure clear communications, attention to details, and a deliverable end product; flag contract expiration dates and confirm contractual milestones
- Unbundle or break down large project into smaller contracts to ensure greater participation of smaller contractors

- Have an owner's technical representative systematically ensure that work performed is per the contract—establish clear accountability, and correct problems before they grow
- Complete the project with accurate invoicing

Summary

Project procurement can be a very important and costly part of a project, and a very important part of project management. The upfront time invested in planning for procurements is extremely important in order to facilitate a procurement process that flows as smoothly and effectively as possible. Project managers should seek to establish procurement relationships between the buyer and sellers where both sides are winners; this is in the interest of everyone involved.

References

Gindlesperger, Willaim, *Supply Chain Management Review*, July 8, 2009.

Smith, Mark. *Supply & Demand-Chain Executive*, October 28, 2003.

About the Authors

Dr. Eldon R. Larsen is and AIChE Fellow and is a professor of engineering at the Marshall University Graduate College in South Charleston, WV. He serves as a Director in the AIChE Management Division board. He has taught project management courses in industrial and academic settings for over 16 years.

Paul Jason Dean works for the state of Kentucky in Transportation and manages projects constantly as part of his daily work.

(171a) Tools and Techniques for Managing Innovation From Beginning to End

Jack Hipple, Innovation-TRIZ, Tampa, FL, **Mukund Karanjikar**, Technology Holding LLC, Salt Lake City, UT and **Eldon R. Larsen**, Marshall University Graduate College, South Charleston, WV

Innovation and its management is the hot topic of the last decade in many industries, including the chemical industry. We have come to realize that Six Sigma and DFSS can eliminate variation in what we know how to do, or plan to do, but cannot by themselves produce innovation.

Innovation needs to start with an idea coupled to a need. Its flesh and bones need to be provided by people and their supporting organizations. Finally, the business and manufacturing concepts behind the innovation must be implemented by smart project management.

This session will review the necessary skills to make this journey successful, including proper idea identification, putting together innovation and project teams, and successful project management.

See more of this Session: R&D Management From Beginning to End
See more of this Group/Topical: Management Division

(171b) Finding the Right Ideas

Mukund Karanjikar, Technology Holding LLC, Salt Lake City, UT

This part of the Managing Innovation workshop will focus on techniques to identify the best opportunities for serious innovatin efforts.

See more of this Session: R&D Management From Beginning to End
See more of this Group/Topical: Management Division

Choosing the Right Team to Innovate

Jack Hipple
Innovation-TRIZ
Tampa, FL
8143-994-9999
www.innovation-triz.com

Abstract

Once focus areas and ideas are selected for an innovation effort, the people who will actually work on an innovation team are chosen. Frequently this is done through a "who's available" and "who knows the most" approach with little regard for personality styles and potential conflicts that can decrease the efficiency of the innovation effort. This portion of the workshop will focus on tools and assessments that are available to assist teams in improving their innovation efficiency.

1. Background and Introduction

The structuring of innovation teams is important. Everyone knows this or at least recognizes its importance. What kind of thought goes into assembling a team to work on a project or new product which has a major component of innovation within it? Well certainly representation from all the various functions within the organization is important. This would usually include research, marketing, manufacturing, legal and patent, as well as customer service and quality. The backgrounds and experiences of these people is important of course, but how often do we spend thinking about the *styles* of the people on the team. Now this is more than *liking* each other. This is about how people interact with each other and how they think about problems and challenges. Now this may in fact be important to the team leader who doesn't want to spend his or her time playing referee. But if it just mean that and nothing more, then it just means peace, not necessarily productivity.

2. Personality Styles

There are many personality style assessment measures that are used in industry to assist individuals in their career growth and in short term feedback, but seldom are such tools used pro-actively in team structuring and formation. In fact, many individuals, when asked what their "type" is (from a previous assessment) often produce a blank stare. They remember some of the "letters", but usually not all of them, nor what they mean. For the purposes of this presentation we will use the widely used Myers Briggs assessment and its four characteristics:

Extroversion/Introversion (E vs. I): This refers to the external communication style of a person. A person who loves to talk with strangers, glad hand at a function vs. someone who would prefer to talk on the phone, curl up with a book, and would not look forward to a large gathering of strangers

Sensing/Intuition (S vs. iN): This describes the difference between individuals who like to take in facts and data for decision making vs. an intuitor who might be just as comfortable with gut feelings and instincts or data that's not obvious at first glance.

Thinking/Feeling (T vs. F): Referring to the difference between some who makes decisions based solely on facts and data vs. concern for the human side and people impact of actions and decisions.

Judging/Perceiving (J vs. P): This refers to the difference between the desire to bring closure to a process and making final plans, critical paths, etc. vs. leaving the door open for possibilities not seen or readily obvious ("not putting all the eggs in one basket").

These styles of people would predict very different approaches to innovation. An ESTJ would dominate a team meeting from a verbal standpoint and only be interested in facts and data and who's going to do what when. An INFP would be relatively quiet (maybe their ideas would not be heard in the team meeting!), their concerns would be about the unsaid things by the customer and how the new process or product might affect the need for overtime on the assembly line and its operators' families. They might also be asking about whether all the ramifications had been thoroughly thought out. There is no right or wrong about any of these styles, but assembling a team without knowing the personalities and preferences is like putting together a baseball team and randomly assigning positions without tryouts and knowing who can pitch or catch or bunt.

3. Problem Solving Style

Another strongly embedded characteristic of people is their approach to problem solving. This can be measured via an assessment instrument known as the Kirton KAI™. This assessment returns a number between 32-160 with a 2σ deviation of 70-110. The lower the KAI™ profile, the more adaptive and structured the thinking and solution generations will be. Comfort within the existing paradigm is very high and there is high discomfort with ideas and solutions outside the existing paradigm. Adaptive KAI people love agendas, goals, objectives, and commitment dates. Their thought process is very externally visible—they explain it to you so you don't have to guess. You can spot one almost immediately by simply looking at someone's desk and seeing how orderly it is! At the other extreme will be individuals who, in a brainstorming session, will throw out ideas without evaluation or assessment (quantity is what is important), and who are not so concerned about plans and deadlines. Lastly, the manner in which they reach a conclusion or an answer to a problem is not so obvious and they usually need to backtrack and explain where they were coming from.

Let's now take a look at how there two primary characteristics of behavior and style preferences can affect team dynamics.

4. Team Dynamics

Let's think a minute about an innovation team that has been assembled to bring on stream a new product developed by a corporation's research department. What if we just assembled the team based on functional needs and paid no attention to the factors we've just reviewed. If half the team was extroverted and half introverted, what mechanism or team leadership is required to make sure that all ideas and concerns are heard? If the team is developing a new product requiring much testing and data analysis and the majority of the members are "N's", how likely is it that all the proper data will be gathered? If there's a contractual commitment data with penalties involved for late delivery, would it not be very important to make sure that there is good representation of "J's" on the planning team? If there is concern about broad patent coverage, it would be good if there were a few "N's" in the room to think about possibilities, not just today's products.

Let's think about a team that's challenged with a new competitive product. The adaptive KAI's on the team will be thinking about negotiating, cross licensing, and other "in the current structure" ideas. Highly innovative KAI™ personalities will be thinking about alternative products and patent circumvention. Are either of these "right" or "wrong"? Of course not! The point is that with only one type of person in the room, we only get one type of idea. And without knowing the profiles and openly discussing these in a team setting, it is not possible to be sure that we have all the right people in the room for the innovation challenge we have at hand.

™KAI is a registered treademark of M.J. Kirton

(171d) Manging Your Innovation Project

Eldon R. Larsen, Marshall University Graduate College, South Charleston, WV

Innovation projects are more complex than standard implementation projects as they frequently involve technology and partnerships outside the normal range of corporate activities. This portion of the workshop will focus on the necessary project management steps and skills to manage innovation related projects.

See more of this Session: R&D Management From Beginning to End
See more of this Group/Topical: Management Division

(171e) Discussion on Innovation Management

Jack Hipple, Innovation-TRIZ, Tampa, FL, **Mukund Karanjikar**, Technology Holding LLC, Salt Lake City, UT and **Eldon R. Larsen**, Marshall University Graduate College, South Charleston, WV

Participants will submit questions and discussion points for the presenters to respond to and comment upon from their experiences.

See more of this Session: R&D Management From Beginning to End
See more of this Group/Topical: Management Division

(229a) Why University-Industry Partnerships Matter

Anthony M. Boccanfuso, The National Academies, University-Industry Demonstration Partnership, Washington, DC

Increasing attention is being paid to university-industry research partnerships since they are perceived to be critical to advancing the discovery to innovation process and competitiveness.

While several advantages of these type of collaborations are generally agreed upon, there are numerous challenges and barriers and this session will discuss a number of topics:

* The advantages of engaging in university-industry partnerships

* The challenges faced in developing effective collaborations

* The various perspectives of companies, universities and the faculty and industry scientists and engineers who pursue these type of partnerships

* Suggestions for identifying appropriate partners

* The impact of government agencies (state and federal) and foundations on these partnerships.

and finally, an update on the activities of the National Academies' University-Industry Demonstration Partnership and the projects being pursued to achieve its mission.

See more of this Session: Managing Technology Collaborations I - Academia, Industry and Government
See more of this Group/Topical: Management Division

(229b) IP Challenges in University-Industry Collaborations

Ken Horton, Gore School of Business, Westminster College, Salt Lake City, UT

Universities and their Industry partners generally have the same business goals when they collaborate on research and development. One of the exceptions to this general rule, though, arises when the partners discuss the topic of the intellectual property (IP), who owns it, and how it is used. Some of these challenges are a result of the traditional business models that are used in the collaboration. These challenges can be overcome by utilizing new business models in the collaboration that treat the IP--and how it is used--differently.

See more of this Session: Managing Technology Collaborations I - Academia, Industry and Government
See more of this Group/Topical: Management Division

(229c) Managing Multilateral Technology Collaborations at the Institute for Collaborative Biotechnologies

David H. Gay, Institute for Collaborative Biotechnologies, University of California, Santa Barbara, Santa Barbara, CA

The Institute for Collaborative Biotechnologies is a uniquely interdisciplinary research and development alliance. Led by the University of California, Santa Barbara, in collaboration with MIT and Caltech, together with partners from the Army and industry, the ICB transforms biological inspiration into technological innovation. The presentation will address opportunities, challenges, progress achieved and lessons learned in managing multilateral technology collaborations between the Army, industry and academia.

See more of this Session: Managing Technology Collaborations I - Academia, Industry and Government
See more of this Group/Topical: Management Division

Opening Stem Cell Research and Development: A Policy Proposal for the Management of Data, Intellectual Property, and Ethics

Krishanu Saha[1*], David E. Winickoff[2*], and Gregory D. Graff[3*]

[1] Whitehead Institute for Biomedical Research, Cambridge, MA 02142

[2] Department of Environmental Science, Policy, & Management, Univ. of California, Berkeley, Berkeley, CA 94720

[3] Department of Agricultural and Resource Economics, Colorado State Univ., Fort Collins, CO 80523

[*] Authors contributed equally to this work.

Abstract

The need to strike a better balance between free sharing and proprietary and regulatory restraint in the life sciences is becoming an important policy concern in the fields of health policy, law, and bioethics. Here we explore these issues in the field of stem cell research. Expanded funding for stem cell research and development holds unique promise for advancing medicine and human welfare. However, technical, proprietary, and regulatory conditions are not ideal: closed information, congested intellectual property, and regulatory complexity are likely to slow innovation, skew attention towards large markets, and prevent ethical settlement. Accordingly, we propose a new form of collective action among funding agencies, scientific journals, and research institutions to centralize data, cell lines, and technologies and to design and disseminate research tools. The goal is more efficient technology exchange within the research community, orientation of product development towards pressing public health needs, and promotion of 'best ethics' in research and development.

Three Pervasive Problems in Stem Cell R&D

Within each of three policy domains current conditions present serious problems for the pace of innovation, the distribution of resulting health benefits, and the public accountability of stem cell research. Moreover, these problems interact and are mutually compounding across domains.

Exchange of Data and Materials

Enhanced capacities to produce, manage, and disseminate data through new information technologies have fueled debates about the manner and timing of data release in the sciences [1]. Traditional norms around sharing research inputs have run headlong into the desire of scientists and their institutions to protect their interests in potentially valuable data, materials, and research results. Practices of delay and secrecy have resulted from the increased relevance of university life sciences to commercial markets and private investors [2, 3]. One mark of change is the proliferation of material transfer agreements between universities [2, 4]. Recent studies suggest that data withholding is becoming common and problematic [5].

Intellectual Property Rights

Following patenting trends elsewhere in the life sciences, scientists and their institutions have been claiming the essential building blocks of stem cell research as private assets. In the specific case of stem cell lines, the first and most obvious problem concerning materials arose from the combination of the Bush Administration's restrictive funding policies and the commanding patent position of the Wisconsin Alumni Research Foundation [6, 7]. Recent analyses show a significant rate of accumulation of new patents over stem cell technologies [8-11], with problematic implications for downstream innovation [12-14].

Bioethics and Regulations

As if technical and proprietary complexities were not enough, disagreement over ethics has given rise to two major problems for efficient and accountable governance of this work. First, in the U.S. federal moratorium on new human embryonic stem cell (hESC) lines has resulted in a vacuum both of funding and oversight. Even as private and state-funded research moves ahead, a national framework is lacking apart from some 2005 guidelines from the National Academy of Sciences. Second, the resulting state programs have proliferated a patchwork of state regulations that differ, sometimes only slightly, on a range of issues. To amplify matters, similar differences occur globally across national regulatory regimes.

Towards an Integrative Approach

Scientists and policymakers have, of course, identified each of these issues, and important initiatives in stem cell research have been undertaken. These efforts should be applauded, but, across the board, they remain inadequate on two counts. First, none of them goes far enough to solve the particular problem it targets. Second, by targeting technical, IP, or regulatory problems in isolation these efforts neglect the important interconnections between these domains. Overlooking interaction between the different domains involves both a conceptual and a practical error. The conceptual error is to ignore the profound ways in which these domains are mutually constitutive categories. Norms and practices of sharing data and materials are simultaneously issues of property and of ethics. Questions of sharing data and materials raise important questions of property which, in turn, raise questions of ethics. To separate these questions is to perform a conceptual purification that, at a minimum, prevents optimal solutions. Yet, the overlap is not merely conceptual, it is practical. As the recently encountered problems with some of the NIH hESC lines illustrate, ethical accountability can be promoted only to the extent that provenance characteristics and data about cell lines are well known and shared [15, 16]. Efficient progress of ethically accountable stem cell research requires considerations that span the complexities and bottlenecks across all three domains. Not only would an integrative approach be more efficient, given the interdependencies, an integrative approach may be the only way to solve some of the individual problems.

The Logic of Collective Action

The problems discussed above have not been adequately addressed largely because they are classic collective action problems, in which a diverse set of actors share common interests yet no one individual actor's incentives are sufficient to overcome the inertia of inaction. Mobilization requires leadership in the form of coordination and initial investment. Sufficient conditions for collective action have not quite yet emerged in any one of the three domains, let alone across them. And, given the currently decentralized pattern of funding and regulation in the U.S.—despite the articulation of clear interests among many parties—there is no clear leader. Under these circumstances, it is necessary to motivate *many, heterogeneous* potential actors through the use of various carrots and sticks. We argue that collective action with a push from major funding agencies can be a basis for opening up stem cell research in the face of multiple compounding constraints. It is precisely the major funding agencies, such as CIRM or the NIH, that have important roles to play in supporting and enhancing common-access resources in stem cell research. Such an "opening up" would result in a more efficient exchange of data, and with further action, sufficient access to research tools within the stem cell research community. We propose four specific actions for stem cell policy:

1. Lead the development of practical common norms

First, major funding agencies could convene interested segments of the research community to construct a practical normative framework for how stem cell materials and data should be managed and circulated within the community. Participation in the efforts above indicate that this group is heterogeneous and large—including scientists, lawyers, ethicists, state funding agencies, nonprofit agencies, universities, and industry. Leadership from major funding agencies is needed to structure this discussion in finding common ground. This commonly developed normative framework should seek to redress the negative effects spanning the technical, proprietary, and ethical domains.

2. Require access in research contracts

Second, major funding agencies should use funding contracts to require the sharing of research data and materials. By many estimates, stem cell lines and the fundamental protocols of cellular reprogramming and differentiation constitute common resources for the research community that are expected to drive broad progress in biomedical research. Thus, cell lines contain the hallmarks of what have previously been called "community resource projects" [17, 18]. Accordingly major funding agencies could classify the development of new human stem cell lines as a "community resource project". A key consequence of invoking this norm would be to include sharp stipulations in research contracts concerning access to stem cell lines and their associated technologies. The stipulations need not use a simple "open access" model, but could be structured in a variety of creative ways that would increase their flow while respecting the private rights and claims of inventors and their institutions [19].

3. Build a common data resource

Third, major funding agencies should support a common data resource that encompasses all of the necessary technical, IP, and ethics information that will be used by its funded scientists, and hopefully, by the community of stem cell scientists globally. The lack of success in this area to date suggests that generating and sustaining support and interest for such an initiative will require new carrots and sticks, and complementary efforts among major funders.

4. Foster further collection action

Fourth, major funding agencies should support further efforts to coordinate, design, build, and disseminate research tools and platform technologies to alleviate the most important technical, IP, and ethical bottlenecks facing the field. These first three actions can construct the foundation for further collective action in which research institutions themselves, and not the government, could to take the lead. With sufficient data available, consortia of interested research institutions could perform systematic analyses of key stem cell lines, research tools, or other platform technologies where some combination of technical, proprietary, and ethical bottlenecks have stymied work at the participating institutions and/or their commercial partners. Analysis of particular constraints for a particular cell line, research tool, or other platform technology is a time-intensive task that, if centralized and done once for many, would have significant efficiency gains for scientists, technology transfer offices, and ethics review boards alike. Informed by the results of those common analyses the consortia could go so far as to designing new technologies that work around the most important technical, IP, or ethical bottlenecks. These new technologies could then be developed and made available for both research use and commercial development of downstream applications through collective licensing or "protected commons" strategies—such as open source or patent pooling. The constituent components would be ethically vetted and chosen to meet regulatory standards across multiple jurisdictions. Such creative initiatives have taken root in the fields of agricultural biotechnology and synthetic biology [19-22], and could prove useful for stem cell research [23].

CONCLUSIONS

The field of stem cell research is full of promise. Yet, those who seek today to realize that promise must wade through a morass of data and material transfer issues, navigate or negotiate numerous intellectual property claims, and stand in peril of violating ethical or formal regulatory requirements. Major funding agencies can and should do more with its leadership than simply fund more research, taking a more active role in expanding the institutional underpinnings of a science commons that is under strain. Such leadership could catalyze fundamental normative and institutional changes that would expedite the exchange of technologies within the research community, orient investments over a wider range of public health needs, and promote a more ethically sound advance of stem cell R&D.

REFERENCES

1. Rai, A.K. and R.S. Eisenberg, *Harnessing and Sharing the Benefits of State Sponsored Research.* Berkeley Technology Law Journal, 2006. **21**: p. 1187-1213.

2. Merrill, S.A. and A.M. Mazza, *Reaping the Benefits of Genomic and Proteomic Research: Intellectual Property Rights, Innovation, and Public Health.* 2006: National Academy Press.

3. Merges, R.P., *Property Rights Theory and the Commons: the Case of Scientific Research.* Social Philosophy and Policy, 1996. **13**: p. 145-167.

4. Ku, K. and J. Henderson, *The MTA?rip it up and start again?* Nature Biotechnology, 2007. **25**: p. 721-721.

5. Caulfield, T., et al., *Patents, commercialization and the Canadian stem cell research community.* Regen Med, 2008. **3**(4): p. 483-96.

6. Wadman, M., *Licensing fees slow advance of stem cells.* Nature, 2005. **435**: p. 272-273.

7. Loring, J.F. and C. Campbell, *Intellectual Property and Human Embryonic Stem Cell Research.* Science, 2006: p. 1716-1717.

8. Esmond, R.W., R.A. Schwartzman, and T.J. Ebersole, *Stem Cells: The Patent Landscape.* Intellectual Property & Technology Law Journal, 2006. **18**(1): p. 1-4.

9. Glanzel, W., *Stem Cells - Analysis of an emerging domain of scientific and technological endeavour.* Steunpunt O&O Statistieken, K.U. Leuven, 2004.

10. Campbell, D., M. Noiseux, and G. Cote, *Potential for Stem Cells Science and Technology in Canada: Great Promises and Challenges.* Science-Metrix report, 2004: p. 63-63.

11. Holden, C., *Wisconsin Stem Cell Patents Upheld.* Science, 2008. **319**(5870): p. 1602b-1603-1602b-1603.

12. Scheinfeld, R.C. and P.H. Bagley, *The current state of embryonic stem cell patents.* New York Law Journal, 2001. **226**.

13. O'Connor, S.M., *Intellectual Property Rights and Stem Cell Research: Who Owns the Medical Breakthroughs?* New England Law Review, 2005. **39**.

14. Simkin, M., *How will intellectual property factors affect the commercialization of stem cell therapeutics.* 2005: BioEurope.

15. Streiffer, R., *Informed Consent and Federal Funding for Stem Cell Research.* Hastings Center Report, 2008. **May-June 2008**: p. 40-47(8)-40-47(8).

16. Baker, M., *Consent issues restrict stem-cell use.* Nature, 2008. **454**(7204): p. 556-556.

17. Wellcome Trust. *Sharing data from large-scale biological research projects: a system of tripartite responsibility. Wellcome Trust Meeting Report, January 14-15, 2003, Fort Lauderdale, FL.* 2003. Fort Lauderdale, FL.

18. Human Genome Program. *Summary of principles agreed at the first international strategy meeting on human genome sequencing.* 1996. Bermuda.

19. Winickoff, D.E., K. Saha, and G. Graff, *Opening stem cell research and development: A policy proposal for the management of data, intellectual property, and ethics.* Yale Journal of Health Policy, Law, and Ethics, 2009. **IX**(Winter).

20. Atkinson, R.C., et al., *Intellectual property rights. Public sector collaboration for agricultural IP management.* Science, 2003. **301**(5630): p. 174-5.

21. Rai, A. and J. Boyle, *Synthetic Biology: Caught between Property Rights, the Public Domain, and the Commons.* PLoS Biol, 2007. **5**(3): p. e58-e58.

22. Porceddu, M.B.C., et al., *Constructive Approaches to Intellectual Property Complexity in Today?s Agricultural Technology World.* Plant Mol. Breed, 2007. **5**: p. 294-295.

23. Winickoff, D.E., K. Saha, and G. Graff, *Opening stem cell research and development: A policy proposal for the management of data, intellectual property, and ethics.* Yale Journal of Health Policy, Law, and Ethics, 2008. **VIII**(Winter).

(287a) University Technology Transfer and Corporate Relations

Leon Przybyla, Technology Transfer, Brigham Young University, Provo, UT

Technology Transfer between universities and corporations bears two very distinct points of view. The emphasis in this presentation will be on the university's point of view. BYU is a private institution and places its primary focus on education. Nevertheless, BYU ranks first in invention disclosures, patent applications, licenses and options, and start-ups per research dollar according to the 2008 AUTM survey. As many as 15-20 start-ups could be spun-out this year, double 2008. BYU is not funded by any of our state's economic incentive funds like USTAR or Fund of Funds, yet it contributes greatly to the economic growth of our state. Universities receive a bad reputation for driving hard bargains and being greedy without receiving proper credit for economic growth stimulus. This begs the opportunity to talk about some of these misconceptions and misunderstandings. It can be argued that universities are essentially places for education not job creation. Many professors believe that all their inventions should be given freely to the world without any formal technology transfer interference. Licensing agreements are a fundamental element of tech transfer. Experience has taught that licensees are leery of any terms in the agreement that may hinder business opportunities. On the other hand, the interests of the university must be protected so general counsel is not fighting unnecessary business law suits. Misunderstandings that occur in licensing negotiations include but are not limited to liability, warranty of technology, termination, ownership of improvements, milestones, and product development versus research. The goals of universities and corporations are surprising similar. Both are trying to maximize return, do future research, make sales, find start-up funding and assistance, and find customers. The best deals are win/win contracts, and finding common ground is essential to success.

See more of this Session: Managing Technology Collaborations II - Academia, Industry and Government
See more of this Group/Topical: Management Division

A Dynamic Ecosystem Approach to Renewable Energy Commercialization
AIChE Annual Meeting, November 2010

Li Han Chan & Nicola Kerslake
Nevada Institute for Renewable Energy Commercialization (NIREC)

NIREC is a clean energy technology accelerator that commercializes renewable energy and energy efficiency technologies. Its experience in developing such technology leads it to believe that progress is greatest where a complex and specific ecosystem is actively created for each innovation. The constraints inherent in renewable energy ecosystems can be eased by diverting public sector funding from research to commercialization activities, by altering incentive structures for researchers within universities and national laboratories and ultimately, by lowering investors' risk perceptions.

The Nevada Institute for Renewable Energy Commercialization (NIREC) is a clean energy technology accelerator that commercializes renewable energy and energy efficiency technologies developed by Nevada and California universities and early stage companies nationwide. It was founded in 2007 with the backing of industry, government and financial partners. The company utilizes a technology pull model; it invests through a request for proposal (RFP) process which funds principal investigators' (PIs) development of pre-commercial technologies. Typically, three projects are selected in each bi-annual RFP round, with the aid of a Technology Commercialization Advisory Board that consists of representatives from industry as well as academic institutions, and potential end-use customers. Each project receives up to $100,000 in funding for technology development, in addition to the services of an experienced entrepreneur-in-residence (EIR) for a one year engagement.

Through the management of these projects, NIREC has come to understand the technology commercialization process as one which entails the creation of a dynamic ecosystem unique to each project, owing to their individual circumstances and the need to circumvent inadequacies in the broader commercialization infrastructure. In this paper, we begin with a discussion of the literature that has influenced our thinking, outline our approach to technology commercialization, and conclude by suggesting improvements in clean energy commercialization infrastructure.

Our data sources are threefold; the eminent and long standing body of academic research on technology commercialization, NIREC's marketplace experience and interviews with technology transfer office (TTO) staff, entrepreneurs, researchers and angel investors.

Literature Review

NIREC's thinking with regard to commercialization has been influenced by academic studies that emphasize cross-border comparisons and a cultural approach to the issues facing ecosystem participants. We highlight two bodies of work as examples.

Bekkers, Gilsing and van der Steen (2006) attribute the success of university intellectual property (IP) spin offs to the contributions of national, sectoral, regional and managerial institutions. Using a comparison between the Dutch and US models to illustrate the impact of structural differences, Bekkers et al argue that national and sectoral institutions determine a university spin off's creation and that TTO policy and management determine its longer run chances of survival. Briefly, Bekkers et al compare the US favorably to the Netherlands as follows:

- *National Law and Policy.* The authors views US legislation – especially the Bayh-Dole Act of 1980 – as key to the better returns to research and development spend seen in the US over recent years (see exhibit 3). Lacking such legislation, the Dutch system is viewed as less able to encourage spin off activity.

- *Sectoral Characteristics.* The availability of technology for transfer is determined by the available sectoral knowledge base. The authors differentiate between formal or codified knowledge transfer and the knowledge transfer that takes place in informal settings, called tacit knowledge. Further, they describe some technologies being developed using one skill set or discipline (mono-disciplinary) whereas others require several skill sets (multi-disciplinary). The transfer of codified knowledge is important for the establishment of spin offs, but tacit knowledge is useful for sustenance of the spin off. In this regard, geographic clusters are important as they enable the transfer of tacit knowledge. The easiest transfer method – as it involves the fewest participants – is in codified mono-disciplinary areas such as biotechnology, while multi-disciplinary and tacit areas (in which we would include renewable energy) are more challenging. Again the authors believe that the US's strong geographic clusters give it an edge over the Netherlands.
- *TTO Policy.* The majority of US university TTOs have adopted a proactive stance towards technology licensing, whereas Dutch ones have not all recognized the benefits of technology transfer, nor are they incentivized to do so (Wintjes et al, 2002).
- *Spin Off Management.* Dutch spin-outs favor a tools or services business model which involves consulting or selling research results; 65-70% of spin offs use this model. The PI generally remains with the firm and this can have mixed results depending on the individuals' management skills. In the US, professional management is preferred and more spin outs focus on creating 'blockbuster products'. The latter approach has a higher failure rate, but surviving firms have stronger long run growth rates.

Overall, the authors conclude that Dutch technology transfer mechanisms could be improved by reference to US ones.

Elsewhere, Roux (2006) views the technology transfer process as being impeded by cultural differences between ecosystem participants (researchers, policy makers, resource managers). As the author describes his thesis:

> *"Sustainable ecosystem management relies on a diverse and multi-faceted knowledge systems in which techniques are continuously updated to reflect current understanding and needs"*

He proposes a "co-production of knowledge" rather than the more commonplace technology or knowledge transfer. Wilson (1998) uses the term 'consilience' to explain a combination of knowledge across disciplines to define a common groundwork. He argues that diverse disciplines, such as anthropology, physics and the arts, are intrinsically unified and that knowledge should be viewed as a process rather than as an entity.

Finally, Roux outlines three methods of pushing knowledge across the divide between scientists and managers; involve end users in the knowledge creation process, improve scientist credibility and package scientific information in more manager-friendly language.

NIRECâ€™s Dynamic Ecosystem Approach

In its experience over the past two years, NIREC has found that the majority of academic commercialization theories assume a static ecosystem and that learnings from individual empirical case studies are replicable. Our experience has been that the complexity of inter-relationships between ecosystem participants, along with rapid changes in the wider commercialization infrastructure, renders a prescriptive approach ineffective. Moreover, the number of variables in any given ecosystem is so large that, in our experience, empirical learning from one ecosystem can be used only as a reference point, not as a blueprint, for problem solving in another.

In reaching this view, we make a number of underlying assumptions:

- Demand creates its own supply
- Industry incumbents do not demand innovative renewable energy products
- The US taxpayer has invested a substantial amount in basic research, and is not seeing a sufficient return on this investment. As is shown in exhibit 3, the US ranks well behind Korea, Japan, China, Russia and Germany in the number of resident patent filings per $ million of research and development spend.
- Many technology transfer models exist to aid in commercializing technology
- Part of market failure is that ecosystem participants do not communicate their needs to one another in a common language
- The most effective common language to accelerate clean energy commercialization â€“ a price signal in the form of a carbon price or similar â€“ is absent in the US and likely to remain so for some time

We delineate between ecosystem variables â€“ the factors which can be controlled or influenced by NIRECâ€™s short run choices for an innovation â€“ and the wider ecosystem infrastructure, where this denotes factors which cannot be controlled by NIREC either in the short run or, in some cases, at all. This viewpoint is summarized below.

Exhibit 1: Dynamic Ecosystem Overview

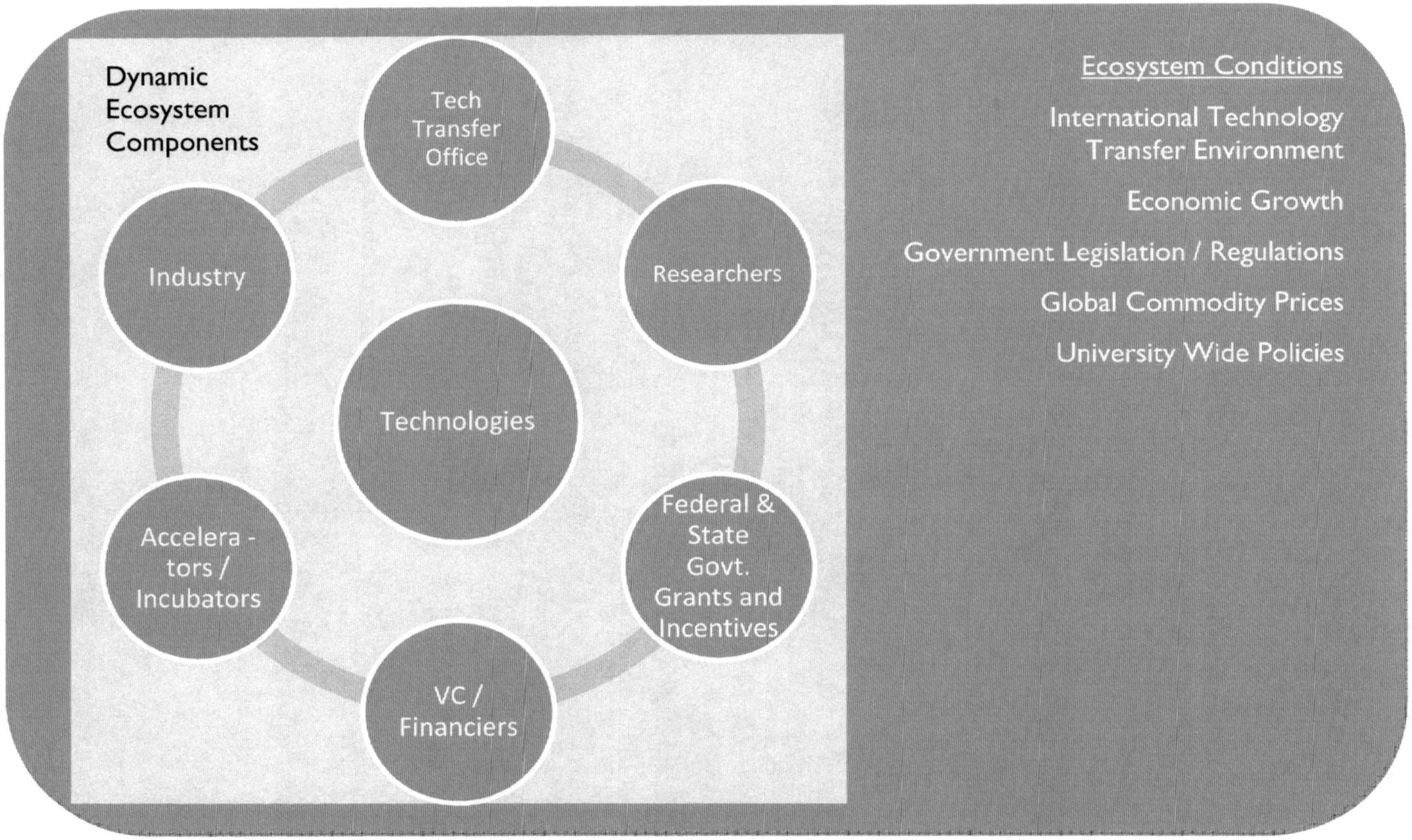

In assembling a commercialization team for any given project, we begin by identifying the motivations and aspirations of each ecosystem participant, some examples of which are:

Exhibit 2 : Motivations & Aspirations of Ecosystem Participants

Ecosystem Participant	Potential Motivations & Aspirations
Researchers	Long-term career aspirations of researchers are overwhelmingly to remain in an academic career, either as a researcher (34%) or as a lecturer (26%)[1] Researchers are more likely to act as 'connectors' between ecosystem participants than are others[2] Even the most entrepreneurial researchers may wish to act as consultants to the start ups based on their technology, rather than derailing their tenure track
Tech Transfer Office	Licensing technology is not directly articulated as a priority for most; 39% of TTOs view 'faculty service' as their top priority, a further 35% cite 'translating research results'[3] There is persistence in spin off activity; O'Shea found that, of the top 20 spin off universities in 80-94, 11 were still in the top 20 in 95-01[4] Licensing fees only account for around 3% of research spend[5], so it is not a strong financial priority for most TTOs
Industry	Incumbent energy industries (e.g. utilities and oil companies) are regulated in such a way that does not encourage radical innovation Energy sector spends only 0.3% of revenues on R&D vs. 10% for pharmaceuticals[6]
Accelerators/ Incubators	As an example, NIREC tasks its EIRs with raising incremental project funding by the end of its year-long engagement: all projects have achieved this goal
VC / Financiers	VCs seek investments with low capital intensity and swift return profiles; these are few and far between in renewable energy Investors are notoriously wary of unproven technologies: 66% of angel investors rate assessment of emerging technology as a high or very high priority, vs. 43% of existing technologies[7]
Federal & State Govt	Motivations are typically to encourage a combination of economic growth, job creation and social development U. S. political will is lacking to develop policies to implement energy price signals required for radical private sector innovation

[1] Academic Research Careers for Scotland data
[2] 'Networks of Inventors and Academics in France, Italy and Sweden', Lissoni et al, 2007
[3] Brandeis University Office of Technology Licensing data, 2009
[4] 'Entrepreneurial Orientation, Technology Transfer and Spin Off Performance of US Universities', O'Shea, 2005
[5] 2.9% in 2004, per AUTM data
[6] NIREC estimates
[7] Survey for Environmental Venture Technology Assessment Program, Center for Venture Research, 1996

Exhibit 3 : International Comparisons and Time Series of Return on Research Spend

No of Resident Patent Filings per $mn R&D Expenditure

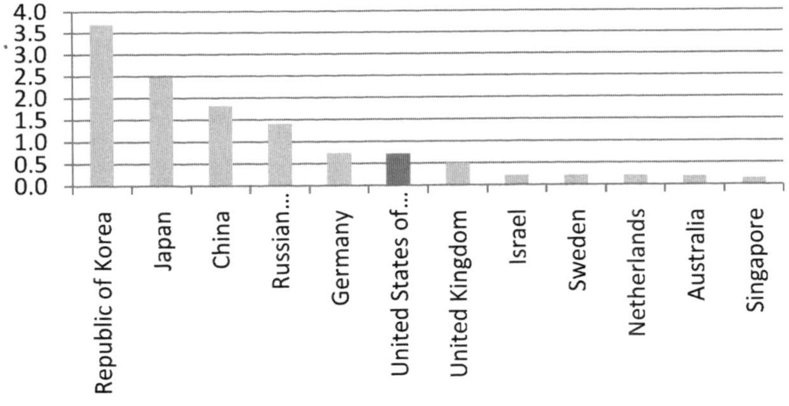

Index of Resident Patent Filings /$mn R&D Spend, 2007 vs. 1998

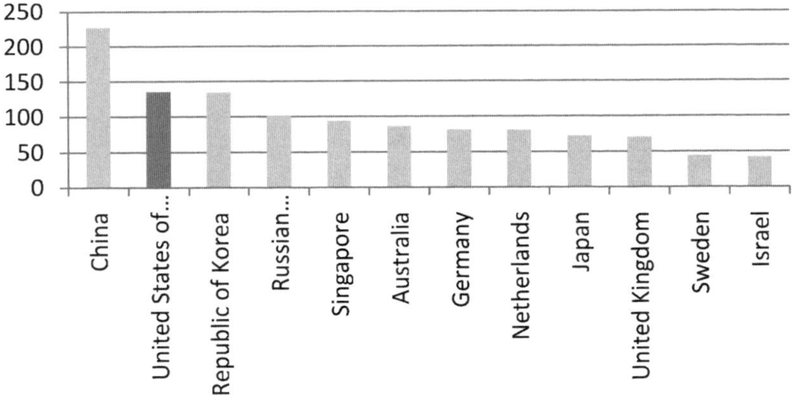

Resident Patent Filings per $bn Gross Domestic Product

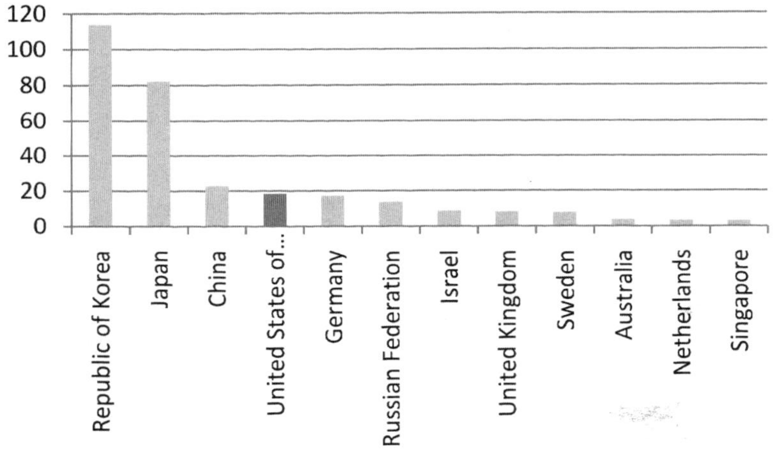

Index of Resident Patent Filings / $bn GDP, 2007 vs. 1998

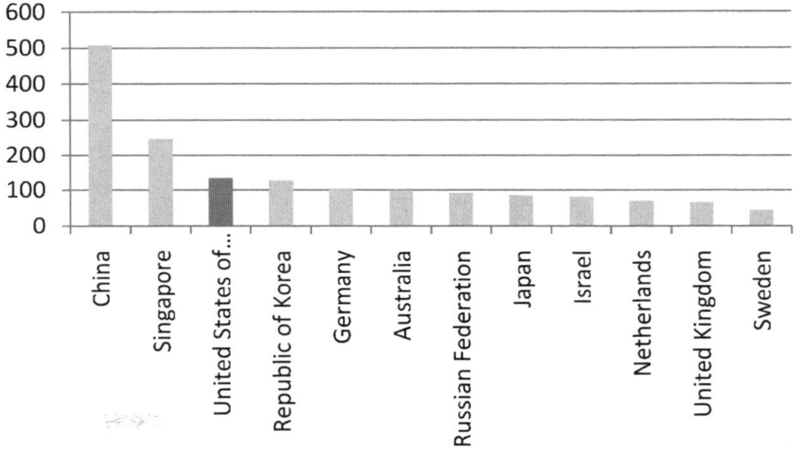

Source: World Intellectual Property Association
Change data is an index of 2007 vs. 1998, aside from Australia in top right hand chart, which is for 2007 vs. 1997.

Once the attitudes and capabilities of each ecosystem participant has been determined, it is possible to assemble a project team that best capitalizes on the ecosystem's strengths and shore up weaknesses; in a typical scenario, two or three participants would be key to development with the remainder playing only a minor role. Notably, these roles often shift over the course of an engagement as the project's environment and development stage alters.

Ecosystem Conditions

In any case, we are maximizing the project's potential within ecosystem conditions that are more constraining for renewable energy technologies than for other sectors, such as pharmaceuticals and information technology. In our view, renewable energy faces more severe challenges in financing and technology development areas than do more established sectors.

Competitive Pressures

We believe that the renewable energy industry faces more intense competitive pressures in its technology development than did industries such as software at the same stage of development.

Though the US is mid-range for all technology development (see exhibit 3), it lags Japan in renewable energy inventions, as is shown in exhibit 4, meaning fewer commercialization opportunities for US entrepreneurs. Additionally, from an early stage in its development, the renewable energy industry has been subject to international technology transfer pressures. In a 2009 study, CERNA found that the rate of international technology transfer was comparable in the relatively immature renewable energy industry, compared to those of more established industries[8].

Renewable energy also remains a relatively small commercialization field in the US; most university spin offs are in the biotech and information technology spaces[9], so there are a smaller number of commercialization opportunities. Energy does not behave like other sectors such as software: there has been no radical "game changing" technology in a generation.

Exhibit 4:Top 10 Renewables Inventors, With Avg % Of Total Inventions (1998 - 2003)

Country	Rank	Avg % of world inventions	Most important technology classes
Japan	1	40.8%	All technologies
USA	2	12.8%	Wind, solar, hydro, methane, buildings
Germany	3	12.7%	Biomass, Ocean, Waste, CCS, wind, solar
China	4	5.8%	Cement, geothermal, solar, hydro, methane
South Korea	5	4.6%	Lighting, ocean, hydro, biomass, cement
Russia	6	4.2%	Geothermal, cement, hydro, CCS, ocean
France	7	2.4%	Cement, CCS, buildings, biomass, hydro
UK	8	1.9%	Ocean, biomass, wind, methane
Canada	9	1.5%	Hydro, wind, CCS, ocean
Brazil	10	1.1%	Ocean, building

Source: CERNA, from 'Invention and Transfer of Climate Change Mitigation Technologies on a Global Scale: A Study Drawing on Patent Data' of December 2008

[8] 'Invention and Transfer of Climate Change Mitigation Technologies on a Global Scale: A Study Drawing on Patent Data', CERNA, December 2008

[9] 'The Performance of University Spin Offs', Zhang, 2008

Financing Pressures

Commercialization activities are hampered by the need to obtain funding for both pilot stage ('Valley of Death') and first commercial stage ('Mountain of Death') project implementation. Biotechnology projects are aided through these stages by an industry that thrives on new product development, while software businesses require little capital. By contrast, renewable energy is part of an industry (energy / utilities) that has few incentives to innovate and is disadvantaged, in the eyes of venture capitalists and other investors, by its high capital requirements and relatively lengthy time to market.

Moreover, renewable energy is primarily a field that is focused on applied research, rather than the basic research which receives the bulk of federal funding. Across all research, there is a roughly 35:1 disparity in federal funding for basic research versus the funding for applied research intended to lead to commercialization.[10]

Recommendations

Given these constraints, our recommendations for easing of ecosystem constraints are:

- Legislation to create price signals that reflect the societal cost of fossil fuel emissions (e.g. price on carbon)
- Increase levels of government funding toward technology development and demonstration (scale-up) to assist in overcoming the 'Valley of Death' and 'Mountain of Death' issues
- Fund commercialization infrastructure to ease private sector financing hurdles through the efficient use of capital to de-risk technology and validate business models
- Incentivizing researchers by offering state, federal or university grant funding for commercialization, not just for basic research
- Reward commercialization activities as part of a career path or tenure track
- Increase the number and intensity of programs whereby federal or state government acts as the first customer for a business. A demonstrated customer with field results dramatically improves the chances of obtaining incremental funding.

Conclusions

In conclusion, NIREC views each renewable energy technology commercialization innovation as a specific ecosystem, consisting of dynamic ecosystem components that can be directly influenced, and ecosystem conditions, that are static in the short run. Renewable energy ecosystem conditions are more challenging than those in information technology and pharmaceutical industries. These challenges could be partially alleviated by diverting government and university funding to commercialization activities and by altering researcher incentive structures and ultimately, investor risk perceptions.

[10] AUTM figures

Bibliography

Aggarwal, Aradhna. "Technology Policies and Acquisition of Technological Capabilities in the Industrial Sector: A Comparative Analysis of the Indian and Korean Experiences." *Science Technology & Society* 6, no. 2 (2001): 255 -304.

Asheim, B. T, and A. Isaksen. "Regional innovation systems: the integration of local 'sticky'and global 'ubiquitous' knowledge." *The Journal of Technology Transfer* 27, no. 1 (2002): 77–86.

Becker, Barbara, and Oliver Gassmann. "Corporate Incubators: Industrial R&D and What Universities can Learn from them." *The Journal of Technology Transfer* 31, no. 4 (7, 2006): 469-483.

Bekkers, Rudi, Victor Gilsing, and Marianne Steen. "Determining Factors of the Effectiveness of IP-based Spin-offs: Comparing the Netherlands and the US." *The Journal of Technology Transfer* 31, no. 5 (9, 2006): 545-546.

Berger, Martin, and Javier Revilla Diez. "Technological Capabilities and Innovation in Southeast Asia." *Science Technology & Society* 11, no. 1 (March 1, 2006): 109 -148.

Berman, E., and S. Machin. "Skill-biased technology transfer around the world." *Oxford review of economic policy* 16, no. 3 (2000): 12.

Bozeman, B. "Technology transfer and public policy: a review of research and theory." *Research policy* 29, no. 4-5 (2000): 627–656.

Buckley, P. J. "International technology transfer by small and medium sized enterprises." *Small Business Economics* 9, no. 1 (1997): 67–78.

Cardozo, Richard, Alexandre Ardichvili, and Anthony Strauss. "Effectiveness of university technology transfer: an organizational population ecology view of a maturing supplier industry." *The Journal of Technology Transfer* (2, 2010).

Clarysse, Bart, Mike Wright, Andy Lockett, Philippe Mustar, and Mirjam Knockaert. "Academic spin-offs, formal technology transfer and capital raising." *Ind Corp Change* (August 2, 2007): dtm019.

Dechezleprêtre, Antoine, Matthieu Glachant, Ivan Haščič, Nick Johnstone, and Yann Ménière. "Invention and Transfer of Climate Change Mitigation Technologies on a Global Scale: A Study Drawing on Patent Data." *SSRN eLibrary* (November 19, 2009).

———. "Invention and Transfer of Climate Change Mitigation Technologies on a Global Scale: A Study Drawing on Patent Data." *SSRN eLibrary* (November 19, 2009).

Dechezleprêtre, Antoine, Matthieu Glachant, and Yann Ménière. "The Clean Development Mechanism and the international diffusion of technologies: An empirical study." *Energy Policy* 36, no. 4 (April 2008): 1273-1283.

Dimov, Dimo, and Gordon Murray. "Determinants of the Incidence and Scale of Seed Capital Investments by Venture Capital Firms." *Small Business Economics* 30, no. 2 (3, 2007): 127-152.

Djokovic, Djordje, and Vangelis Souitaris. "Spinouts from academic institutions: a literature review with suggestions for further research." *The Journal of Technology Transfer* 33, no. 3 (9, 2006): 225-247.

Fuller, Douglas B. "How law, politics and transnational networks affect technology entrepreneurship: Explaining divergent venture capital investing strategies in China." *Asia Pacific Journal of Management* 27, no. 3 (6, 2009): 445-459.

Gibson Raymond, W., and V. David. "Key variables in technology transfer: a field-study based empirical analysis." *Journal of engineering and Technology Management* 8, no. 3-4 (1991): 287–312.

Glachant, Matthieu, Antoine Dechezleprêtre, and Yann Ménière. "The Clean Development Mechanism and the International Diffusion of Technologies: An Empirical Study." *SSRN eLibrary* (December 2007).

Goldfarb, B., and M. Henrekson. "Bottom-up versus top-down policies towards the commercialization of university intellectual property." *Research policy* 32, no. 4 (2003): 639–658.

Gorschek, Tony, Per Garre, Stig Larsson, and Claes Wohlin. "A Model for Technology Transfer in Practice." *IEEE Softw.* 23, no. 6 (2006): 88-95.

Hackett, Sean M., and David M. Dilts. "A Systematic Review of Business Incubation Research." *The Journal of Technology Transfer* 29, no. 1 (1, 2004): 55-82.

Hakanson, Lars, and Robert Nobel. "Organizational characteristics and reverse technology transfer," August 19, 2010.

Handbook of Bioentrepreneurship. Vol. 4. New York, NY: Springer New York, 2008.

Henderson, R., A. B Jaffe, and M. Trajtenberg. "Universities as a source of commercial technology: A detailed analysis of university patenting, 1965-1988." *Review of Economics and Statistics* 80, no. 1 (1998): 119–127.

Kedia, B. L, and R. S Bhagat. "Cultural constraints on transfer of technology across nations: Implications for research in international and comparative management." *Academy of Management Review* 13, no. 4 (1988): 559–571.

Klein, Raphael, Uzi Haan, and Albert I. Goldberg. "Overcoming obstacles encountered on the way to commercialize university IP." *The Journal of Technology Transfer* (11, 2009).

Leitch, Claire M., and Richard T. Harrison. "Maximising the potential of university spin-outs: the development of second-order commercialisation activities." *R and D Management* 35, no. 3 (6, 2005): 257-272.

Libaers, Dirk, Martin Meyer, and Aldo Geuna. "The Role of University Spinout Companies in an Emerging Technology: The Case of Nanotechnology." *The Journal of Technology Transfer* 31, no. 4 (7, 2006): 443-450.

Lissoni, Francesco, and Bulat Sanditov. "Networks of Inventors and Academics in France, Italy and Sweden: evidence from the Keins Database." KNOWLEDGE-BASED ENTREPRENEURSHIP: INNOVATION, NETWORKS AND SYSTEMS (2006).

Mansfield, E. *Intellectual property protection, foreign direct investment, and technology transfer*. The World Bank, 1994.

Meseri, O., and S. Maital. "A survey analysis of university-technology transfer in Israel: evaluation of projects and determinants of success." *The Journal of Technology Transfer* 26, no. 1 (2001): 115–125.

Mowery, David C., and Joanne E. Oxley. "Inward technology transfer and competitiveness: the role of national innovation systems." *Camb. J. Econ.* 19, no. 1 (February 1, 1995): 67-93.

O'Shea, Rory P., Harveen Chugh, and Thomas J. Allen. "Determinants and consequences of university spinoff activity: a conceptual framework." *The Journal of Technology Transfer* 33, no. 6 (9, 2007): 653-666.

O'shea, R. P, T. J Allen, A. Chevalier, and F. Roche. "Entrepreneurial orientation, technology transfer and spinoff performance of US universities." *Research Policy* 34, no. 7 (2005): 994–1009.

Park, W., and D. Lippoldt. "Technology transfer and the economic implications of the strengthening of intellectual property rights in developing countries." *Document de travail de l'OCDE sur la politique commerciale n°62, OCDE, Paris* (2008).

Phan, Phillip H., and Donald S. Siegel. *The Effectiveness of University Technology Transfer*. Now Publishers Inc, 2006.

QUINN, I. B. "Technology transfer by multinational companies." *Multinational Business Operations: Environmental aspects of operating abroad* (1973): 160.

Roux, D. J, K. H Rogers, H. Biggs, P. J Ashton, and A. Sergeant. "Bridging the science-management divide: moving from unidirectional knowledge transfer to knowledge interfacing and sharing" (2006).

Siegel, Donald S., Reinhilde Veugelers, and Mike Wright. "Technology transfer offices and commercialization of university intellectual property: performance and policy implications." *Oxf Rev Econ Policy* 23, no. 4 (December 1, 2007): 640-660.

Swamidass, Paul M., and Venubabu Vulasa. "Why university inventions rarely produce income? Bottlenecks in university technology transfer." *The Journal of Technology Transfer* 34, no. 4 (12, 2008): 343-363.

Takim, R., R. Omar, and A. H Nawawi. "International Technology Transfer (ITT) Projects and Development of Technological Capabilities in Malaysian Construction Industry: A Conceptual Framework." *Asian Social Science* 4, no. 8 (2009): P38.

Wonglimpiyarat, Jarunee. "Commercialization strategies of technology: lessons from Silicon Valley." *The Journal of Technology Transfer* 35, no. 2 (3, 2009): 225-236.

Wright, Mike, Sue Birley, and Simon Mosey. "Entrepreneurship and University Technology Transfer." *The Journal of Technology Transfer* 29, no. 3/4 (8, 2004): 235-246.

Wu, Weiping. "Managing and incentivizing research commercialization in Chinese Universities." *The Journal of Technology Transfer* 35, no. 2 (3, 2009): 203-224.

Zander, Udo, and Bruce Kogut. "Knowledge and the Speed of the Transfer and Imitation of Organizational Capabilities: An Empirical Test." *Organization Science* 6, no. 1 (February 1995): 76-92.

Zhang, Junfu. "The performance of university spin-offs: an exploratory analysis using venture capital data." *The Journal of Technology Transfer* 34, no. 3 (4, 2008): 255-285.

(287c) Lessons Learned by An Entrepreneurial Minded University Professor On the Creation of New Companies

Bruce Gale, Department of Mechanical Engineering, University of Utah, Salt Lake City, UT

In this presentation, the author will present several lessons learned by the author during the process of spinning out companies based on his research at the University of Utah. The presentation will specifically address 4 questions or areas that might be relevant to academics looking to start a company. First, what should a professor consider before working to spin out a company? Some of these issues include: time commitments, involvement of graduate students, conflict of interest questions, motivation for the spin out company, and what is required to be successful? The author will look at pros and cons of the technology spin out process. Second, what can a University do to assist professors with the commercialization process? This discussion will focus on how the goals of the University and the Technology Transfer Office may not always match up with the goals of the professor, and how these goals can be reconciled to maximize benefits to both parties. Specifically, how can RPT and commercialization coexist? Third, how can the University partner with the spin out company? This question is important if many of these small businesses are to thrive. Small businesses have many challenges, and the assistant of a large University can help. Fourth, how can the small business contribute back to the University. The small business has many ways it can return the favor to the University, including money, research opportunities, and education. Finally, the author will present some personal case studies to help address these questions and hopefully provide some answers to academics considering starting a spin out business.

See more of this Session: Managing Technology Collaborations II - Academia, Industry and Government
See more of this Group/Topical: Management Division

Licensing as a Business Strategy

American Institute of Chemical Engineers

Annual Meeting

November 9, 2010

Steve Ludmerer
Ludmerer & Associates LLC
856-273-9895
steve@ludmerer.com

Steve Ludmerer

- **Member of the Licensing Executives Society with over thirty-five years experience taking new concepts and technologies from concept to multi-million dollar sales: Teflon® II coatings, Reach® toothbrush, Tychem® protective fabrics, Parmod® conductive materials.**
- **In- and Out-Licensed a broad range of technologies including diagnostics, polymers, medical devices, electronic materials, consumer products, and others to companies worldwide. Grew and monetized businesses through angel and venture financing, alliances, licenses and public offerings.**
- **Experience in DuPont and Union Carbide/Dow and as leader of three entrepreneurial companies (immunoassay diagnostics, instrument and laboratory business electronic materials).**
- **Often collaborates with Foresight Science and Technology (www.foresightst.com) for business and technology assessments, licensing strategy and deal advisories. Focus is on advanced materials, nanotechnology, specialty chemicals, medical devices, and consumer products.**

Outline

- What is Licensing
- Why License, When to License
- Finding Licensees
- Choosing Licensees
- Valuing your IP
  - Royalty Rates/Transaction Data
  - Situational Factors
  - Discount Rates
- Example
  - Low Risk
  - High Risk
- Licensing Alternative
- Take Aways
  - Sources
  - Additional Assistance

What is licensing?

Licensing is the transfer of *intangible assets* (intellectual property) from one firm or organization to another.

Intangible Assets include:

- Patents
- Know-How
- Copyrights
- Trademarks
- Trade Secrets
- Other Protectable Assets

Why license (for the Licensor)

- Organization is R&D focus (University, Medical Center, R&D firm, etc.)
- Innovation outside organizational fit, resources
- Infrastructure needed for commercialization
  - Manufacturing
  - Market access – domestic and international
  - Functional staffing
- Focus resources on core development
- Component in the value chain – needs pull through
- Time and dollars to realize potential
- Competitive Landscape requires a family of offerings
- SWOT analysis indicates "gaps" can be "better" filled by others
- Licenses Can be Restricted in Field/Geography

Why license (for the Licensee)

- ☐ Strategic fit
- ☐ An asset that will make money
- ☐ Protectable competitive advantage
- ☐ Sword or Shield

Differentiate from something others can acquire (real estate, skilled scientists, marketers, free agent ballplayers, etc.)

When to license

- Licensor capabilities
  - Licensor lacks capability to advance concept to commercialization
- Value increases
  - Patent status
  - Risk Reduction is reduced
  - Cost/Time to revenue decreases

For example.......

Seven Steps in Business Creation

- ☐ Idea Creation/Generation
- ☐ Scoping/Strategic Fit
- ☐ Business Assessment
- ☐ Development
- ☐ Business Plan
- ☐ Pre-Launch
- ☐ Commercialization

Business Value Evolves as Risks/Unknowns are Understood

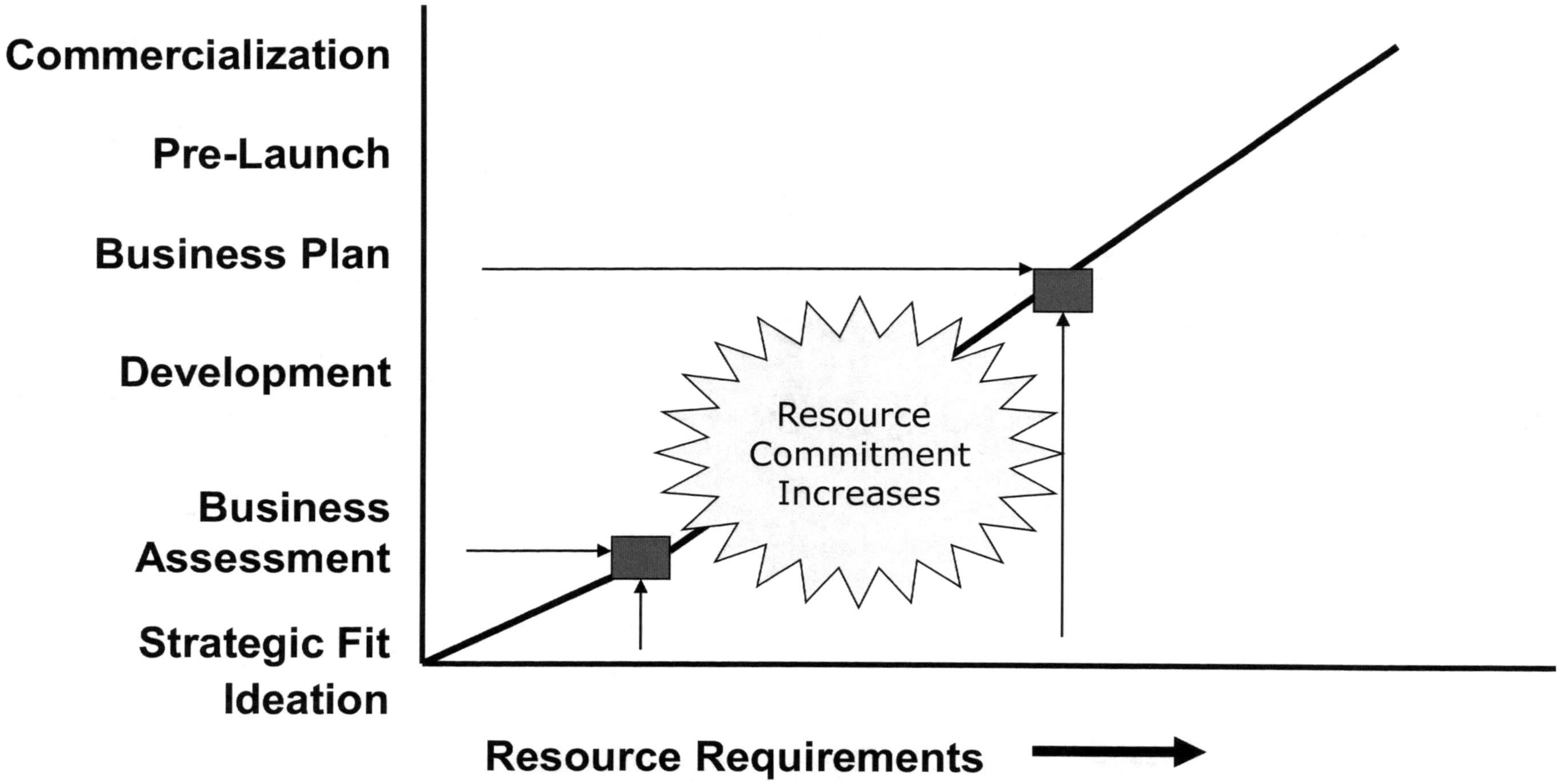

License Value Increases with Technology Readiness Level (TRL)

- TRL 1 Basic principles observed and reported
- TRL 2 Technology concept and/or application formulated
- TRL 3 Analytical and experimental critical function proof-of-concept
- TRL 4 Component/Subsystem/Product (CSP) validation in laboratory environment
- TRL 5 CSP validation in relevant environments (e.g. structural or environmental testing)
- TRL 6 CSP prototype demonstration in a relevant end environment (value chain).
- TRL 7 CSP prototype demonstration in an operational environment (pilot production; beta testing)
- TRL 8 CSP "mission/customer qualified" through test and demonstration in an operational environment
- TRL 9 CSP proven through successful practical use.

Patent Increases in Value

- ☐ Idea Concept
- ☐ Provisional Patent Filed
- ☐ Product Viability Established
- ☐ Domestic Application Filed
- ☐ WIPO Application Filed
- ☐ Patent Granted
- ☐ Invention Demonstration in Use
- ☐ Made, Used and Sold Commercially

Methods to Value Intellectual Property

- Cost Basis
  - No longer used; Sunk cost is irrelevant
- Income Basis
  - How much of the income is attributable to the licensed IP
  - EBIT generated from the licensed IP (versus w/o the IP) less cost of capital for plant and equipment, personnel, marketing, etc.
  - Difficult for licensor to know/estimate
- Revenue Basis
  - Similar IP and industry
  - Industry specific royalty rate experience
  - Reliable disclosure of license agreements
  - Application of situational differences

Finding Licensees

- Analysis of Competition
  - Products
  - Patents
  - Research
- Analysis of Market Space
  - Industry Trade Associations
  - Industry Conferences and Expositions
  - Magazine and periodicals
  - Companies in the market space with other products

Choosing a Licensee

- Strategic Fit
  - Current competitors
  - Potential market entrants
  - Buyers, suppliers, former competitors
- Experience commercializing licensed technology
- Compelling reason for adoption
- Clear and persuasive ability to achieve objective
- Open dialogue toward a win-win deal

Typical Industry Royalty Rates

Royalty Rates for In-Licenisng by Industry							
Industry	0-2%	2-5%	5-10%	10-15%	15-20%	20-25%	>25%
Aerospace	50.0%	50.0%					
Automotive	52.5%	45.0%	2.5%				
Chemical	16.5%	58.1%	24.3%	0.8%	0.4%		
Computer	62.5%	31.3%	6.3%				
Electronics		50.0%	25.0%	25.0%			
Energy		66.7%					
Food/Consumer		100.0%					
General Mfg.	45.0%	28.6%	12.1%	14.3%			
Governmetn/University	25.0%	25.0%	50.0%				
Health Care	3.3%	51.7%	45.0%				
Pharmaceuticals	23.6%	32.1%	29.3%	12.5%	1.1%	0.7%	0.7%
Telecommunications	40.0%	37.3%	23.6%				

Source: NJSBDC Presentation December 8, 2004

Transaction Data - RoyaltySource.com

Industry	Average	Median	Max	Min	Count
Chemicals	4.7%	4.3%	25.0%	0.1%	78
Internet (incl software)	11.8%	8.8%	50.0%	0.3%	88
Telecom (excl Media)	4.9%	4.5%	15.5%	0.4%	73
Consumer Gds, Rtl & Leis	5.5%	5.0%	28.0%	0.1%	98
Media & Entertainment	9.1%	5.0%	50.0%	2.0%	25
Food Processing	3.2%	2.8%	10.0%	0.3%	38
Medical/Health Products	6.1%	5.0%	77.0%	0.1%	376
Pharma & Biotech	7.0%	5.0%	50.0%	0.0%	458
Energy & Environment	5.0%	5.0%	20.0%	1.0%	107
Machines/Tools	5.2%	4.5%	25.0%	0.5%	90
Automotive	4.3%	3.5%	15.0%	0.5%	59
Electrical & Electronics	4.2%	4.0%	15.0%	0.5%	139
Semiconductors	4.3%	3.0%	30.0%	0.0%	75
Computers & Office Equip	5.3%	4.0%	25.0%	0.2%	73
Software	11.5%	6.8%	70.0%	0.0%	147
Industry Summary	6.40%	4.80%			1,924

Source: NJSBDC Presentation December 8, 2004

Situational Factors Impacting Royalty Rates

- ❑ Significance of the Invention
- ❑ Maturity of Technology (TRL)
- ❑ Level of IP Protection
- ❑ Competitive Landscape

- ❑ Platform Technology
- ❑ Field of Use/Geography of Use
- ❑ Exclusivity/ Non-Exclusivity

- ❑ Upfront Payment
- ❑ Minimum Royalty
- ❑ Milestone Payments
- ❑ Duration of Agreement
- ❑ Time to Commercial Marketing/Sales

- ❑ Patent Maintenance and Enforcement Burden
- ❑ Ancillary Services (R&D, Training, Support, etc.)

Typical Discount Rates

Risk Level	Approximate Discount Rate (%)	Description
Risk Free	10 - 18	Existing product, high demand, building more of the same
Very Low Risk	15 - 20	New well understood technology for an existing product
Low Risk	20 - 30	New features, well understood technology into an existing market
Moderate Risk	25 - 35	New product, well understood technology into market with competition
High Risk	30 - 40	New product, not well understood technology into an existing market.
Very High Risk	35 - 45	New product, new technology, new market
Extremely High Risk	50 - 70	New company, unproven technology, new market

Source: Richard Razgaitis, "Early-Stage Technologies: Valuation and Pricing" John Wiley & Son 1999.

An Example

Low Risk Technology License

License Terms	
Royalty rate on gross sales	3%
License Issue Fee(s) ($000)	$100
License Maintenance Fee	$0

Timing	
First year of analysis -->	2010
First year of Market Sales -->	2013

	Discount Rates									
Year	1	2	3	4	5	6	7	8	9	10
Year Ending	2010	2011	2012	2013	2014	2015	2016	2017	2018	2019
Inflation	1.0%	2.0%	2.0%	2.0%	2.0%	2.0%	2.0%	2.0%	2.0%	2.0%
Discount Rate	20.0%	20.0%	20.0%	20.0%	20.0%	20.0%	20.0%	20.0%	20.0%	20.0%
Total Discount for NPV	21.0%	22.0%	22.0%	22.0%	22.0%	22.0%	22.0%	22.0%	22.0%	22.0%

Invention Addressing $200MM Market

STARTING ASSUMPTIONS - MARKET		
Total Market Size for Upcoming Year ($000)		$200,000
Market Growth Rate	yrs 1-4	8.0%
	yrs 5-8	5.0%
	yrs 9-10	4.0%

	Sales Projection By Year									
Year Year Ending	1 2010	2 2011	3 2012	4 2013	5 2014	6 2015	7 2016	8 2017	9 2018	10 2019
Total Market Forecast ($000s)	$200,000	$216,000	$233,280	$251,942	$264,540	$277,766	$291,655	$306,238	$318,487	$331,227
Licensee Market Share				2%	8%	15%	30%	40%	45%	50%
Licensee Gross Sales Forecast ($000s)	$0	$0	$0	$5,039	$21,163	$41,665	$87,496	$122,495	$143,319	$165,613

Low Risk Scenario

Licensee Revenues										
Year	1	2	3	4	5	6	7	8	9	10
Year Ending	2010	2011	2012	2013	2014	2015	2016	2017	2018	2019
Licensee Gross Sales Forecast ($000)	$0	$0	$0	$5,039	$21,163	$41,665	$87,496	$122,495	$143,319	$165,613
NPV of revenues ($000)	$0	$0	$0	$2,275	$7,830	$12,636	$21,751	$24,960	$23,937	$22,672
Cumulative NPV of Revenues	0	$0	$0	$2,275	$10,105	$22,741	$44,492	$69,451	$93,388	**$116,060**

Licensor Revenues										
Year	1	2	3	4	5	6	7	8	9	10
Year Ending	2010	2011	2012	2013	2014	2015	2016	2017	2018	2019
License Issue Fee ($000)	$100									
Subtotal	$100	$0	$0	$0	$0	$0	$0	$0	$0	$0
Royalty on gross sales ($000)	0	0	0	151	635	1,250	2,625	3,675	4,300	4,968
Total Licensor Revenue ($000s)	$100	$0	$0	$151	$635	$1,250	$2,625	$3,675	$4,300	$4,968
Cumulative royalty revenue ($000)	$100	$100	$100	$251	$886	$2,136	$4,761	$8,436	$12,735	**$17,704**
NPV of royalty revenue ($000)	$83	$0	$0	$68	$235	$379	$653	$749	$718	$680
Cumulative NPV of royalty revenue ($000)	$83	$83	$83	$151	$386	$765	$1,417	$2,166	$2,884	**$3,564**

New Market, New Technology, New Product

License Terms	
Royalty rate on gross sales	3%
License Issue Fee(s) ($000)	$100
License Maintenance Fee	$0

Timing	
First year of analysis -->	2010
First year of Market Sales -->	2013

	Discount Rates									
Year Year Ending	1 2010	2 2011	3 2012	4 2013	5 2014	6 2015	7 2016	8 2017	9 2018	10 2019
Inflation	1.0%	2.0%	2.0%	2.0%	2.0%	2.0%	2.0%	2.0%	2.0%	2.0%
Discount Rate	40.0%	40.0%	40.0%	40.0%	40.0%	40.0%	40.0%	40.0%	40.0%	40.0%
Total Discount for NPV	41.0%	42.0%	42.0%	42.0%	42.0%	42.0%	42.0%	42.0%	42.0%	42.0%

High Risk Scenario

Licensee Revenues										
Year	1	2	3	4	5	6	7	8	9	10
Year Ending	2010	2011	2012	2013	2014	2015	2016	2017	2018	2019
Licensee Gross Sales Forecast ($000)	$0	$0	$0	$5,039	$21,163	$41,665	$87,496	$122,495	$143,319	$165,613
NPV of revenues ($000)	$0	$0	$0	$1,239	$3,666	$5,082	$7,516	$7,410	$6,105	$4,968
Cumulative NPV of Revenues $(000)	0	$0	$0	$1,239	$4,905	$9,987	$17,503	$24,913	$31,018	$35,986

Licensor Revenues										
Year	1	2	3	4	5	6	7	8	9	10
Year Ending	2010	2011	2012	2013	2014	2015	2016	2017	2018	2019
License Issue Fee ($000)	$100									
Subtotal	$100	$0	$0	$0	$0	$0	$0	$0	$0	$0
Royalty on gross sales ($000)	0	0	0	151	635	1,250	2,625	3,675	4,300	4,968
Total Licensor Revenue ($000s)	$100	$0	$0	$151	$635	$1,250	$2,625	$3,675	$4,300	$4,968
Cumulative royalty revenue ($000)	$100	$100	$100	$251	$886	$2,136	$4,761	$8,436	$12,735	$17,704
NPV of royalty revenue ($000)	$71	$0	$0	$37	$110	$152	$225	$222	$183	$149
Cumulative NPV of royalty revenue ($000)	$71	$71	$71	$108	$218	$371	$596	$818	$1,001	$1,151

Comparison

	Lower Risk	**Higher Risk**
NPV of Total Revenue (000)	$116,000	$36,000
Cumulative Royalty Revenue (000)	$17,700	$17,700
Cumulative NPV of License (000)	$3,564	$1,151

Comparison vs. Venture Investment

	Lower Risk	**Higher Risk**
NPV of Total Revenue (000)	$116,000	$36,000
Basis for Venture Investors – NPV of earnings (000)	$~ 23,200 (est 20%)	$~7,200 (est. 20%)
Cumulative NPV of License (000)	$3,564	$1,151

Takeaways

- Licensing can be an attractive strategy to monetize an innovation
- Licensing can be core or ancillary strategy
- Licensing helps validate patents
- Establishing and defending royalty rates and discount rates is an important consideration in your negotiations
- High risk, high investment scenarios are often attractive for licensing

Helpful Sources

- **Free Sites**
  - Web Search sites (Google - Advanced Search; Google Scholar; Site:.xxx, etc.)
  - Licensing Executives Society - www.usa-canada.les.org
  - SEC Filings - www.sec.gov/edgar.shtml
  - U.S. Patent & Trademark Office - www.uspto.gov
  - Industry Sites
    - ThomasNet - www.thomasnet.com
    - GlobalSpec - www.Globalspec.com
    - PubMed - http://www.ncbi.nlm.nih.gov/sites/myncbi/searches/?db=
    - Energy Information Administration - http://www.eia.gov /

- **Fee for Service Sites**
  - Delphion IP Network - www.delphion.com
  - Frost & Sullivan - www.frost.com
  - High Beam - www.highbeam.com
  - Hoovers - www.hoovers.com
  - Royalty Stat www.royaltystat.com
  - Royalty Source http://royaltysource.com/

- **Ludmerer & Associates LLC / Foresight Science & Technology**
  - T2+2™ Technology Transfer Tools and Training
  - Business Plans, Assessments, Market Overviews, Royalty Rates, Transaction Compilations, Pipeline Partners, Venture Capital Sources, Angel Sources

Contacts

Thank You!

Steve@Ludmerer.com

www.ludmerer.com

(33a) Succeeding through Alliances: A Business Strategy

Anthony Nickens, Ceramatec, Inc., Salt Lake City, UT

In today's world of complex innovation challenges, alliances are indispensible. Sooner an entity can realize this, faster it can progress. Formation of technology commercialization alliances has been important at Ceramatec, a high technology company engaged in solving tough R&D challenges. Ceramatec has succeeded at spinning off companies, licensing technologies out and in-house commercialization of several technologies due to having alliances at the very core of business strategy. The presentation will discuss case studies and share lessons learned from several years experience.

See more of this Session: Managing Innovation
See more of this Group/Topical: Management Division

The Soft Side of Innovation: What Engineers Need to Know and Do for Corporate Success

Jeff Lindsay, Innovation Edge, Neenah, WI
jlindsay@innovation.com

Abstract

Chemical engineers often find that excellent problem solving skills and creativity do not translate into success inside a corporation. Some grow frustrated when technically sound solutions and clever innovations fail to have the impact they expected. Some grow disenchanted, concluding that it's all about "politics" instead of logic. In reality, innovation success involves technical expertise plus a network of relationships in order to drive adoption of the innovation. Failure to understand the "soft side of innovation" can contribute to innovation fatigue and unnecessary frustration. This presentation will consider the intangible elements that need to be understood for innovation success, including five key steps for success.

The Problem

Engineers are technically trained and skilled in solving problems–problems that involve mass, energy, chemical reactions, liquid flow, and other non-humanoid materials and products. Engineers often come into corporations, find exciting opportunities for innovation, and then become disappointed and frustrated when unexpected forces seem to quash the concept. The journey of engineers as innovators can be a lonely and aggravating one.

Learning from a Cosmic Analogy: Dark Energy and Dark Matter

In the past decade or so, scientists have been astounded to discover that the universe that we can see represents only a tiny fraction of the matter and energy that governs the cosmos. Based on the motion of stars and galaxies, strange "dark matter" must be present, increasing the gravitational tug on celestial bodies more than can be accounted for by visible matter. Further, based on the surprising discovered that the universe is expanding, not contracting under its own gravitational pull as expected, scientists have proposed that a strange, repulsive "dark energy" fills the cosmos countering gravity. The combined effect of these unseen entities, dark energy and dark matter, are so great, that they account for 96% of the matter and energy of the universe. In other words, the visible universe that we used to think was everything actually is only a tiny fraction of what is out there. What we can see in the "cosmic org chart" accounts for only 4% of what really influences the cosmos.

It's that way in the business world. too. Companies can create tidy org charts and draft neat process maps to describe how they work, but the unseen reality outside the visible systems may be what really dominates operations. Increasingly, experts in knowledge management are learning that easily overlooked and often invisible *intangibles* can dominate corporate value and performance. Numerous intangible transactions may be essential to the success of a company, including casual information sharing between trusted friends, helpful exchanges of tips and best practices between employees or between external partners and internal employees, or loyalty that is gained when people are included in decision making. The invisible linkages and hard-to-observe exchanges in a company's internal an external ecosystems may be the real engines of value creation, regardless of what is on a process map or workstream. By not understanding the value of such intangibles, corporations can easily break key linkages and crush subtle engines of value creation.

Many companies focus on their "value chains" - a term popularized by Michael Porter in his seminal 1985 work, *Competitive Advantage*. The value chain describes the linear chain of events as materials and products move from sourcing through manufacturing and out to the market. It is a highly useful paradigm for manufacturing and was highly applicable to much of the economy in the era when Porter was doing his research. But since that time, the explosion of the knowledge economy has changed the way we work and create value. Verna Allee, a revolutionary expert in knowledge management, has detailed the move from the value chain to modern ecosystems and Value Networks in her book, *The Future of Knowledge: Increasing Prosperity through Value Networks* (Burlington, MA: Elsevier Science, 2003). Verna Allee and Associates have introduced a clever, methodical tool called Value Network Analysis for analyzing and visualizing the transactions of intangibles and tangibles that affect a business. That's beyond the scope of this presentation, but it helps us realize that many of the intangibles that dominate the real world of business can be visualized, revealing bold new insights into how things work

This is Lesson 1 for engineers:

1. **Relationships are the dark energy and dark matter of the corporate universe. They matter more than org charts, titles, and budgets.**

Real success is not about brown-nosing or gaining the favor of one powerful person. It's about filling your work and your life with healthy relationships in which you build ties, communicate, gain loyalty and trust, do extra-mile favors for others, and nurture good will.. Sharing information and helping others is one way to do this.

Here is a corollary important for building the relationships that can bring success:

2. **Look past titles, salaries, and org charts to appreciate and work effectively with those around you.**

Learning to treat each person as important is the only safe and responsible path. Apart from the moral imperatives for kindness to all, there are also practical reasons why one should never put too much weight on the "visible mass" like titles that an individual has. Remember, official appearances and titles don't describe how things really get done. An employee must engage in some corporate astronomy to understand what really moves the planetary bodies and how to be effective. Understanding the agendas, needs, and, sadly, politics of various corporate groups and individuals is part of that work.

Many times key players have a lot of dark matter and thus a lot of influence that others can't see. These can include seemingly low-ranking people who might be easy for some people to ignore or even treat rudely. Administrative assistants, receptionists, sales staff, and many others without flashy titles can wield surprising influence. Learn how to work effectively with the people that can help you get things done. Always treat others, regardless of title, with respect. This includes not just those in the company, but your partners, vendors, suppliers, and other outsiders you work with. Again, these relationships can matter far more than you imagine.

Other Lessons

3. **It's all about sales. You must sell.**

Some engineers are tempted to look down on sales staff and consider their job of selling as something beneath the lofty ranks of engineering and science. That attitude will limit engineers, for in reality, success in a corporation is always about sales. No project moves forward unless it can be sold convincingly to those who must pay for it–management. Innovation does not happen without selling. A great invention does not become an innovation until it is adopted and used by others, and that requires selling at many levels.

In fact, almost every day engineers will have the opportunity or need to sell their work, their ideas, and their skills to others, whether it's their team, their supervisor, other teams in the corporation, or outsiders. It's especially important if you are trying to pursue innovation success.

As for the work of actually selling products to outside customers or partners, never look down on those who are exclusively engaged in the art of selling. They are the engines that drive your company. Smart engineers should learn from them to better understand markets, needs, opportunities, problems, etc., as well as just learning how to sell. Relationships with sales staff can only be helpful in your career.

4. **Inoculate yourself against the corporate antibodies designed to kill innovation.**

Enthusiastic engineers love to solve problems creatively and can be tremendous sources of innovation, a blessing that everyone needs and no one wants–or at least that's how it can seem when numerous internal barriers to innovation seem to stand in the way of innovators. This is a topic addressed more fully in *Conquering Innovation Fatigue* by Jeffrey Dean Lindsay, Cheryl Perkins, and Mukund Karanjikar (John Wiley & Sons, 2009). Whether it's politically-motivated "Not Invented Here" syndrome or well-intended corporate policies that stifle efforts to move innovative projects forward, most corporations seem to have large quantities of anti-innovation antibodies in their system that will strike when you are doing your best to take the company forward. Be intellectually prepared and understand that the reasons for the barriers are not likely to be mere stupidity or evil, but genuine concerns and risks that others see. Learning to see from the perspective of others can help you anticipate and overcome concerns more effectively. Again, building relationships is critical here for getting the information, help and perspectives you need. Innovation success is possible, but you must be prepared for the antibodies and have plans that help you go forward, when possible, in spite of these barriers.

5. **Be able to be held accountable for every action.**

The hidden networks of relationships entwined in an organization can result in swift punishment for stupid deeds or unkind words. They can result in others learning of gaps in behavior. For innovation success, you need to avoid these pitfalls and have the corporation and its people on your side as much as possible, not questioning your motives or accusing you of wrongdoing. There are problems enough for innovators as it is.

For every email you send, imagine a panel of critics later reviewing it to see if you were acting appropriately. When you browse the web at work, imagine you boss looking over your shoulder, because they may very well be doing that. When you chat on a corporate phone, never speak in ways that would embarrass you if someone were listening in. Corporations are not designed to be fair, and even the best of intentions and purest of actions may be misunderstood, but at all costs don't do stupid things that can harm your career. Be ready to be held accountable for every action and avoid the temptation to take moral shortcuts. For best results, strive for best behavior all the time. That's one step toward maintaining the healthy relationships you need to find innovation success in a corporation.

For further insights into innovation success, see http://Innovationedge.com/blog and http://InnovationFatigue.com.

(33c) New Chemical Introduction at Lubrizol

Frank van Lier, The Lubrizol Corporation, Wickliffe, OH

Lubrizol uses a staged New Chemical Introduction Process (NCIP) for managing new component development. This multi stage process has proven its value in many years of application and recently underwent an upgrade to refresh, refocus and revitalize the program. An explanation of the multi-stage process will be provided along with examples of the challenges and keys to success.

See more of this Session: Managing Innovation
See more of this Group/Topical: Management Division

(33d) Improvisation Techniques for Chemical Process Innovation

Peter J. Ludovice, School of Chemical & Biomolecular Engineering, Georgia Institute of Technology, Atlanta, GA, Lew Lefton, School of Mathematics, Georgia Institute of Technology, Atlanta, GA and Richard Catrambone, School of Psychology, Georgia Institute of Technology, Atlanta, GA

The importance of creativity in technical innovation has been strongly emphasized by commercial, educational and government organizations. Such organizations include the National Academy of Engineering and the National Science Foundation. While numerous protocols for design and trouble-shooting have been developed, they typically ignore the creative act within the innovation process. This is likely due to the systematic approach typically employed by engineers in developing such processes. While these protocols are effective at organizing the appropriate concepts, constraints and resources to tackle design and trouble-shooting challenges, they do little to catalyze the creative act in innovation. Recognizing that a more chaotic search through idea space can improve innovation, the application of techniques used in humorous improvisation has catalyzed innovation in non-technical areas. The equivalence of humor and innovation has long been recognized by Arthur Koestler and Edward de Bono. However, we hypothesize that the relative lack of success for this approach in technical areas is due to the additional constraints on feasible innovative technical ideas.

Our approach is based on the equivalence of humorous improvisation for searching idea space and stochastic simulation for searching physical state space.[1] Humorous improvisation is used as a random idea generator, just as random number generators can produce random moves in physical state space. However, the state distribution bias used in stochastic simulation is lacking in humorous improvisation and our recent efforts have focused on developing this equivalent biasing in the searching of idea space. Such a biasing would cause our improvisation protocol to slowly evolve to produce unique and unpredictable innovative technical ideas just as physical stochastic simulations evolve to a unique unpredictable state distribution. Such processes, in which simple stochastic elements produce unpredictable complex behavior, are called emergent processes. Simulations are valuable precisely because such unique complex behavior emerges from such simple simulation elements. Sawyer has recognized that creative ideas emerge from simple group improvisation elements in much the same way.[2] Recognizing the emergent nature innovation from improvisation helps in assessing creativity. Because true innovation is a rare event, it is difficult assess. However, the emergent nature of this process means that innovation is not a single random new idea, but an idea that emerges from a scaffold of ideas build by the group. We are planning to assess the effectiveness of this method by measuring the progress of such a scaffold.

This approach has been applied in various exercises with academic and professional engineering groups and feedback has been taken in the form of post-exercise questionnaires and videotaping of the improvisation sessions. The exercises were carried out by the authors who represent a unique combination of experience in science, engineering, humor, improvisation and educational psychology. We will discuss the results of these exercises and characterize what best practices have been derived from these results. The specific application of this improvisation technique to the chemical process industry will also be discussed.

[1] Ludovice, P.; Lefton, L.; Catrambone, R. "Improvisation for Engineering Innovation," Proc. ASEE National Meeting, Louisville, KY, Paper AC-2010-1650, 2010.

[2] Sawyer, R.K. "Group Genius: The Creative Power of Collaboration", Basic Books: New York, 2007.

See more of this Session: Managing Innovation
See more of this Group/Topical: Management Division

(406a) Business Development, IP, and Manufacturing Success: Perspectives From Utah's Manufacturing Extension Partnership

David Sorensen, Manufacturing Extension Partnership of Utah, Orem, UT

The Manufacturing Extension Partnership of Utah has assisted many companies in strengthening their strategy for success and continued growth. We will discuss what it takes to advance your business, including lessons relative to leadership, vision, intellectual property, and coping with changing regulations and policies.

See more of this Session: Intellectual Assets in the Digital Era
See more of this Group/Topical: Computing and Systems Technology Division

(406b) Role of IP In Successful Startups

Michael Alder, Technology Transfer Office, Brigham Young University, Provo, UT

Many AIChE members will be involved with a startup at some point in their career. While the capabilities of the management team is of utmost importance, in numerous cases, the success of the startup also depends on the quality of its intellectual property. In this era, an IP-savvy team can take several steps to secure competitive advantage and realize greater value from the technology, products, or services the company offers. This presentation will draw upon experience with many startups and startup teams and will provide guidance to researchers, business leaders, and future entrepreneurs on how to better prepare for success.

See more of this Session: Intellectual Assets in the Digital Era
See more of this Group/Topical: Computing and Systems Technology Division

(406c) An Introduction to IP Law: The Underpinnings of Intellectual Assets

Ken Horton, Gore School of Business, Westminster College, Salt Lake City, UT

An understanding of the basics of intellectual property law can help chemical engineers in advancing their own research, in evaluating competitive efforts, in building their own business, or in general advancing their career. This presentation will cover some of the key concepts that engineers should know, including the nature of patents, the different kinds of patents (provisional, utility, design), the role of trademarks and copyrights, what it takes to be patentable, and how changes in patent law may affect your career and business.

See more of this Session: Intellectual Assets in the Digital Era
See more of this Group/Topical: Computing and Systems Technology Division

(406d) Cost-Effective Pursuit of IP in a Down Economy

Jonathan Lee, Salt Lake City Office, Advantage Law Group, South Jordan, UT

How does one get the most protection and benefit from intellectual property when the economy is down? How can patents and other forms of intellectual property be obtained in a cost effective manner when budgets are tight? In this presentation, an experienced patent attorney shares insights into cost effective IP with guidance directed to managers, research leaders, inventors, and entrepreneurs.

See more of this Session: Intellectual Assets in the Digital Era
See more of this Group/Topical: Computing and Systems Technology Division

AUTHOR INDEX